Thinking Geographic

This book explains how the concepts of geography can teach young people to think geographically, deeply and ethically.

Thinking Geographically demonstrates how the concepts of place, space, environment and interconnection teach students new ways of perceiving and understanding the world, the concepts of scale and time teach them ways of analysing the world, while the concepts of sustainability and wellbeing show them how to evaluate and reflect on what they observe, and all eight concepts develop their higher order and critical thinking. To further support teachers, this book includes a chapter on how to teach for conceptual understanding, as well as two chapters that illustrate the application of geographical thinking to an understanding of the effects of land cover change and the problem of regional inequality.

Rich with practical examples, this book is an essential resource for geography teachers, whether already teaching or studying to become one, and for those who teach them.

Alaric Maude is a retired associate professor of Geography at Flinders University in Adelaide, South Australia. He has conducted research in Tonga, Indonesia, Malaysia and Australia, was the lead writer for the Australian Geography curriculum and has published articles for geography educators on powerful knowledge, sustainability and geography's 'big ideas'.

Thinking Geographically

A Guide to the Core Concepts for Teachers

Alaric Maude

LONDON AND NEW YORK

Designed cover image: Alaric Maude, 10th Avenue, New York City, viewed from The High Line, 2013.

First published 2024
by Routledge
4 Park Square, Milton Park, Abingdon, Oxon OX14 4RN

and by Routledge
605 Third Avenue, New York, NY 10158

Routledge is an imprint of the Taylor & Francis Group, an informa business

© 2024 Alaric Maude

The right of Alaric Maude to be identified as author of this work has been asserted in accordance with sections 77 and 78 of the Copyright, Designs and Patents Act 1988.

All rights reserved. No part of this book may be reprinted or reproduced or utilised in any form or by any electronic, mechanical, or other means, now known or hereafter invented, including photocopying and recording, or in any information storage or retrieval system, without permission in writing from the publishers.

Trademark notice: Product or corporate names may be trademarks or registered trademarks, and are used only for identification and explanation without intent to infringe.

British Library Cataloguing-in-Publication Data
A catalogue record for this book is available from the British Library

Library of Congress Cataloging-in-Publication Data
Names: Maude, Alaric, author.
Title: Thinking geographically: a guide to the core concepts for teachers/Alaric Maude.
Description: First edition. | New York: Routledge, 2024. | Includes bibliographical references and index. |
Identifiers: LCCN 2023038416 (print) | LCCN 2023038417 (ebook) | ISBN 9781032453767 (hbk) | ISBN 9781032453736 (pbk) | ISBN 9781003376668 (ebk)
Subjects: LCSH: Geography—Study and teaching. | Geography teachers—Training of.
Classification: LCC G73 .M346 2024 (print) | LCC G73 (ebook) | DDC 910.71—dc23/eng/20231018
LC record available at https://lccn.loc.gov/2023038416
LC ebook record available at https://lccn.loc.gov/2023038417

ISBN: 978-1-032-45376-7 (hbk)
ISBN: 978-1-032-45373-6 (pbk)
ISBN: 978-1-003-37666-8 (ebk)

DOI: 10.4324/9781003376668

Typeset in Galliard
by codeMantra

Contents

Figures, tables and boxes

Figures

Tables

Boxes

Acknowledgements

Many people have helped me to think about the ideas in this book and taught me much. If I have omitted you from the list, my apologies. My particular thanks and appreciation are to:

The teachers who formed the core of the writing group for the Australian Geography curriculum over a decade ago and taught me much about geography in schools, especially Malcolm McInerney, Nick Hutchinson and Susan Caldis.

David Lambert, for encouragement and advice for over a decade, and for leading the way (with John Morgan) with their 2010 book *Teaching geography 11–18: A conceptual approach*. This was a book that established conceptual thinking as a core part of geographical education.

Those who read and commented on one or more of the draft chapters: David Lambert, David Marsh, John Lewin, Ron Martin, Grace Healy, Jeana Kriewaldt, Susan Caldis, Paul Batten, Richard Maude, Jennifer Carter, David Hollinsworth, Olga Gostin and Victor Gostin.

Julie Russell for proofreading and editing over half the chapters and learning some geography along the way.

The organisers of the 2015 and 2019 IGU-GCE London Conferences at the Institute of Education, University College London, for providing unexpected challenges that stimulated my thinking and learning.

The many geographers who contributed to *Geography: Shaping Australia's Future*, a report produced by the National Committee for Geographical Sciences of the Australian Academy of Science, which expanded my awareness of what geographers actually do.

The staff at Routledge in Melbourne, for agreeing to publish a book by someone they had never heard of, and helping me to do it.

And Sandy Policansky, for her advice, support, love and endurance through the years of writing.

Alaric Maude
Adelaide
July, 2023

Introduction

Alexander Murphy, an eminent American geographer, starts his book *Geography: Why it matters* (2018) with a case study of the Lake Chad region in Africa. Murphy compares the region now with what it was like in the early 1960s. Since then, the lake has lost 90% of its surface area, the population has more than doubled, far fewer people live by fishing and there are conflicts from the expansion of agriculture into pastoral areas. Furthermore, the rise of Boko Haram and the military response of national governments has displaced over two million people and produced a food crisis and widespread malnutrition.

Murphy's explanation of these changes is an example of geographical thinking. He argues that to explain what has happened requires an understanding of three sets of factors:

- One is the specific environmental and human characteristics of the region (which is a place). These include a decline in rainfall from natural causes, the expansion of irrigated agriculture resulting from population growth (which took water from the lake), poor governance and local ethnic divisions.
- The second set is the interconnections between environmental and human processes. For example, the shrinking of the lake caused an expansion of the tsetse fly population, which affected the cattle on which island communities depended. These communities were then forced to migrate into the territory of other groups, producing conflict.
- The third set is about influences from outside the region. These include the tensions produced by the political boundaries imposed by the colonial powers, the growth of commercial agriculture for European markets (which added to the expansion of irrigation), climate change caused by fossil fuel use in other countries, Islamic ideologies originating in Southwest Asia and Western support for corrupt governments.

Murphy also identifies the use of spatial analysis to better understand what is happening and what could be done to address problems that vary across the region. His discussion illustrates geographical thinking because it applies the

DOI: 10.4324/9781003376668-1

discipline's core concepts of place, space, environment and interconnection to produce a comprehensive explanation of change in the Lake Chad region.

The purpose of this book is to explain how to use these and other core concepts to teach young people to think geographically, deeply and ethically. Thinking geographically is about the ways that geographers perceive the world and the questions they ask, and how these can teach students new ways of thinking. Thinking deeply is about the ability to understand and explain and to use that knowledge to comprehend the present and think into the future. Thinking ethically is about using the concepts of sustainability and human wellbeing to critically evaluate environmental and social situations and trends and to think about alternatives. Combined they give young people a set of powerful intellectual abilities which they can apply to their own lives.

This book has been written primarily for geography teachers, both those studying to become one and those who are already teaching in schools, but it may also help undergraduate students to recognise the ideas that connect the different topics they are studying and that make them geographical. This book might even be of interest to those who are curious to find out if there is more to the subject than maps, the locations of countries and the names of their capitals.

This book begins with a chapter on knowledge—what we can know (ontology) and how we can know it (epistemology)—because your beliefs about these two questions will influence your use of deeper thinking. For example, if you believe that it is not possible to say that one answer is better than another, because both are equally subjective opinions, then you will see no point in using thinking to distinguish between them. On the other hand, if you think that geographers work 'towards building incrementally justified conclusions based on evidence, with a recognition that such understandings must be considered probably partial and open to refutation' (Alexander et al., 2011, p. 60), then you will see the value of the thinking needed to make these conclusions. The chapter also includes a short section on Indigenous knowledges. This is particularly relevant to countries colonised by Europeans, such as the USA, Canada and Australia, where the Indigenous populations still have a vast body of their own environmental and social knowledge.

Chapter 2 is about thinking geographically. It identifies four key concepts which are central to geography—interconnection, place, space and environment—and explains how these influence thinking in the subject, from ways of perceiving and questioning the world to the choice of policies to address environmental, economic and social issues. It also identifies two analytical concepts—scale and time—and two evaluative concepts—sustainability and human wellbeing. The chapter also explains the importance of conceptual thinking, because intellectual development is concept-based, not fact-based. This is an approach that emphasises meaning and reasoning, rather than just knowing facts. It is also an approach that emphasises that geography's core concepts are ideas that have functions, from identifying questions to finding and evaluating answers, that students can apply to an understanding of a wide range of topics. In highlighting these forms of thinking, this book follows the advice of Kenneth Gregory and John Lewin (2021, p. 15), who write that 'we believe that curricula should require holistic and imagined understandings of the value of the big ideas that disciplines bring'. This level of conceptual thinking tends to be neglected in curriculums and textbooks. For example, Margaret Roberts' analysis of seven English GCSE Geography specifications, which prescribe what students in lower secondary schools should learn, found only two that provided a conceptual framework. She argues that if students are to have access to geography's ways of seeing the world, they need to develop an understanding of the subject's 'big ideas', such as place and space, 'which permeate geographical thinking' (Roberts, 2018, p. 32).

Chapters 3–8 examine the eight major concepts identified in Chapter 2. They explain how these concepts can be unpacked into ideas that students can work with and the different functions they can have. They also review some of the current thinking about them in academic geography. Geography educators regularly express concern over the gap between the subject at the secondary and tertiary levels and the absence of some contemporary academic ideas in school geography. However, school subjects are not stripped-down versions of university disciplines, as they have different purposes (Deng, 2007; Stengel, 1997). The discussion in these chapters will suggest which new ways of conceptual thinking in university geography are appropriate for schools and how they can be used. It will also discuss some concepts that may no longer be taught in universities but have a place in schools, because they are still relevant to students' understanding of their world. These substantive chapters are followed by Chapter 9, which covers ways of teaching the conceptual thinking previously explained and draws on advice from the literature on concept-based teaching and learning. Throughout this book, geography is conceived as a single entity, not separated into physical and human, but unified through its core concepts and ways of thinking (cf. Rawding, 2013, p. 104).

Chapters 10 and 11 illustrate the application of geographical thinking to two major contemporary issues. The first is about the human transformation of the surface of the Earth through the replacement of the native vegetation with crops, pastures, tree plantations and urban areas, and its effects on climate, biodiversity, water resources and Indigenous populations. The second is about places which are economically disadvantaged compared with other places within the same country. These topics have been chosen because they examine two major contemporary challenges, one about environmental change and sustainability and the other about economic inequality and human wellbeing, that tend to be neglected in school geography. They are also good illustrations of the use of geographical thinking. As there are different viewpoints within each topic, there is scope for students to debate alternative opinions and the reasoning and evidence supporting them, which will develop a range of thinking abilities.

The geographical literature drawn on in this book is confined to English-language publications because of my own linguistic deficiencies, and the examples used are mainly from the UK, the USA and Australia. Not all of it is by geographers, as psychologists, sociologists, philosophers and others have all had something to say about geography's core concepts. I have also included some illustrations of the thinking of Australian Indigenous cultures, in order to recognise that there are other ways of perceiving and knowing the world than those of Western societies. The Indigenous peoples of Australia have lived there for at least 60,000 years and developed a deep knowledge of the environment and a rich culture. However, there was and is enormous cultural diversity across Indigenous Australia, with about 250–300 distinct languages and perhaps twice as many dialects, and this makes generalising about Indigenous knowledge and ways of thinking difficult.

A note to the reader

This book covers a wide range of material and ideas. If you are a teacher or studying to become one, you will find that some of it is relevant to primary school, some to secondary school, while some themes are probably only suitable for upper secondary school. However, much of it can be simplified for younger students, and all is relevant to a teacher who wants to understand the conceptual foundations of the subject and the many directions in which it can lead. If you are an undergraduate student, this book will give you an introduction to geography's core concepts and to a range of areas of geographical knowledge and research. If you are neither of these, the book will introduce you to ideas and knowledge about the ways that geographers look at the world that may surprise you and that are highly relevant to your lives. Their value is succinctly described in an Australian report on the discipline:

> Geography is distinctive in its emphasis on spatial thinking, its interest in knowledge generated from the study of specific places, and its recognition of the fundamental importance of the environment to human welfare. Its vision is both local and global. It is also marked by an awareness of the interconnections between phenomena and processes both within places and across space, and its fields of study span the natural sciences, social sciences and humanities. In a world in which inequalities within and between places can threaten social cohesion, the pressure of human impacts on the environment is a growing concern, places and people are increasingly interconnected globally, and problems require answers that integrate different fields of knowledge, Geography has much to offer.
>
> (National Committee for Geographical Sciences, 2018, p. 4)

Keeping up-to-date

In this book, there are references to articles in *Teaching Geography* and *Geography*, two journals published by the Geographical Association (GA) in the UK, and in *The Geography Teacher* and *Journal of Geography*, equivalent journals published by the National Council for Geographic Education (NCGE) in the USA. Serious teachers of geography are encouraged to join both their national, state or province geography teachers' association and the GA, which has an extensive range of resources for teachers. Membership of the GA also provides access to both the UK and US journals and to several others on geographical education.

Teachers interested in keeping up with new information can do so with Google Scholar. With the right choice of keywords, this source can identify recent academic articles on what you are teaching, and a growing number of these may be freely available through a link next to the article. Another search method is to use an article you already know that is relevant to your interests, find it in Google Scholar and then look at the publications that have cited it. Some of these may be on the same topic.

References

Alexander, P. A., Dinsmore, D. L., Fox, E., Grossnickle, E. M., Loughlin, S. A., Maggioni, L., Parkinson, M. M., & Winters, F. I. (2011). Higher order thinking and knowledge: Domain-general and domain-specific trends and future directions. In G. Schraw & D. R. Robinson (Eds.), *Assessment of higher order thinking skills* (pp. 47–88). Information Age Publishing.

Deng, Y. (2007). Transforming the subject matter: Examining the intellectual roots of pedagogical content knowledge. *Curriculum Inquiry*, *37*, 279–295. https://doi.org/10.1111/j.1467-873X.2007.00386.x

Gregory, K. J., & Lewin, J. (2021). Big ideas in the geography curriculum: Nature, awareness and need. *Journal of Geography in Higher Education*. https://doi.org/10.1080/03098265.2021.1980867

Murphy, A. B. (2018). *Geography: Why it matters*. Polity Press.

National Committee for Geographical Sciences. (2018). *Geography: Shaping Australia's future*. Australian Academy of Science.

Rawding, C. (2013). *Effective innovation in the secondary geography curriculum: A practical guide*. Routledge.

Roberts, M. (2018). Do the new GCSEs promote 'sound enquiry and investigative approaches' to learning geography? *Geography*, *103*(1), 27–37. https://doi.org/10.1080/00167487.2018.12094032

Stengel, B. (1997). 'Academic discipline' and 'school subject': Contestable curricular concepts. *Journal of Curriculum Studies*, *29*, 585–602. https://doi.org/10.1080/002202797183928

Chapter 1

What can we know and how can we know it?

We all make assumptions about the nature of reality, what we can know about it and how can we find this knowledge. We may be unaware of these assumptions, but they nevertheless exist and influence the questions we ask and the methods we choose to find answers. They influence whether we think there can be a best answer, or only many opinions, and whether we think it is possible to determine causal relationships and make generalisations about geographical knowledge. If you are a teacher, they influence what and how you teach. As Roger Firth writes, 'Different conceptions of knowledge (and truth) imply and encourage different ideals of thinking, learning, teaching and curriculum in geography' (Firth, 2011, p. 59). Consequently, we cannot avoid some discussion of ontology and epistemology, although at a very basic level.

Ontology is about the nature of reality—about whether there is a real world out there, independent of what we think we know about it (a position called foundationalism), or is the world what we construct it to be and not independent of what we think we know about it (a position called anti-foundationalism). Epistemology is about what we can know about this world, how we can find this knowledge and how we can validate it. Jonathan Grix reduces this explanation to two short questions—ontology is about 'what's out there to know' and epistemology is about 'what and how can we know about it' (Grix, 2019, p. 62). This chapter reviews three different types of answers to these questions—positivism, constructivism and realism.[1] Each of these includes many variants and strands, particularly within constructivism, so the discussion will be very general. Much fuller accounts can be found in the suggested readings at the end of the chapter. The chapter also has a brief discussion of Indigenous knowledges, which have other ways of perceiving and relating to the world.

DOI: 10.4324/9781003376668-2

Positivism

Positivism involves the application of scientific methods to explain natural and social phenomena. It can be described as follows:

- Positivism views the world as real and existing independently of our knowledge and perception of it. Reality is perceived as composed of discrete objects (both tangible, such as beaches, and intangible, such as ethnic groups), which have characteristics that are independent of how they are perceived by an observer.
- Our knowledge of this reality is obtained by observation, using our senses, aided by increasingly sophisticated methods such as remote sensing.
- This knowledge is considered to be objective, because it is determined by the external world, and is assumed to be uninfluenced by the beliefs, values and situation of the observer, or by the social and historical context of the time. In other words, different people will make the same observations if they use the same methods, because they are observing the same objective reality. The fundamental importance of observation stems from the founder of positivism, Auguste Comte, whose aim was to separate science, based on observation, from religious explanations of the world, based on theology.
- Positivists also believe the aim of research is to explain the patterns and regularities that have been found through observation, by constructing generalisations and identifying causal relationships. These generalisations and causal relationships can then be applied to explain and predict other similar events or patterns.
- Hypotheses about causal relationships must be rigorously tested by repeated measurements. Critical rationalism added the concept of falsification, developed by Karl Popper. He argued that if repeated testing produces observations that agree with the hypothesis, then its validity is strengthened, but it is not possible to say that it has been proved because there may be situations not yet discovered that will disprove it. The classic example to illustrate this is that for centuries Europeans believed that all swans were white, because they had only ever seen white swans. Their many observations supported their belief, but it took just one observation of black swans in Western Aus tralia, by a Dutch explorer, to disprove it.

There are some obvious problems with a strict positivist approach to knowledge, mainly, but not only, because it ignores the human element in research. These problems include:

- The research questions chosen by researchers are influenced by their personal interests, experience and beliefs, or by their peers, funders and society. The body of knowledge that we have developed therefore reflects their

subjective choices and may contain gaps. The history of human geography includes successive movements to add neglected subjects to research, such as women, ethnic minorities, Indigenous peoples, people who identify as LGBTQ and more-than-humans. As Susan Hanson pointed out some time ago, questions such as 'the geography of everyday life', 'the links between the unpaid work of caring and work in the paid labor force' and 'the relationship of international migration to child care, domestic work, and the sex trade' were not on geography's agenda until women began asking them (Hanson, 2004, p. 719).
- Observation cannot be totally objective and not influenced by the ideas of the researcher. Karl Popper argued that in scientific practice observation follows theory—a researcher has an idea (a theory) about how to explain some part of reality, and develops a method of collecting data to test the idea.
- What researchers observe of reality is constrained by what they already know, as this influences what they notice or look for. Furthermore, 'any knowledge we derive from the five senses is mediated by the concepts we use to analyze it, so there is no way of classifying, or even describing, experience without interpreting it' (Marsh et al., 2018, p. 187).
- The quantitative methods used by positivist researchers may group observations into categories, and these categories are human constructs.
- Positivist approaches produce grand generalisations that may not fit particular places, and places are often the focus of geographical research.

The founder of positivism, Comte, believed that its methods could be applied to the human world, not just the natural, and that laws of human behaviour could be discovered and used to improve society. This belief, that the social world can be studied by the same methods as the natural world, is called naturalism. However, people, businesses, institutions and societies can change their ways of behaving in response to their reflections on their lived experience, or to changes in the world around them (as can animals), and so in human geography explanation and prediction is difficult, and in the opinion of many non-positivist geographers, impossible.

Is some geography positivist?

A strict form of positivism was abandoned by philosophers long ago. However, within the geographical discipline, physical geographers, and some human geographers, are often considered to be positivists by those who reject this approach, although the latter's construction of what positivism means is generally out-of-date and stereotyped. Many physical geographers could be called positivists because they believe that there is a reality to be discovered that is independent of human thought; that observation and measurement are the ways to discover that reality; that research can develop generalisations and laws about that reality; and that falsification is an appropriate way to test hypotheses. On the other hand, physical geographers are also aware of the criticisms of positivism outlined above, recognise that their knowledge is a human construct of reality and have a far more nuanced approach to research than human geographers generally give them credit for. Rob Inkpen and Graham Wilson, for example, caution against a too simplistic view of how physical geographers (and probably many human geographers) actually investigate the real world. In their book on philosophy in physical geography, they criticise the view that physical geography should follow the positivist scientific approaches of disciplines like physics:

> This view tends to perpetuate the idea that there is an objective reality from which we draw real knowledge. Reality is there—we are clever and extract real data from it to find out what reality is really like. This view is one that is difficult to sustain once you start the practice of actually doing physical geography. At this point, another view of how reality is understood may become more appropriate: the view of reality as constructed by a dialogue. … The dialogue metaphor highlights the negotiated nature of reality. Reality is not just a thing to be probed and made to give up pre-existing secrets. Reality enters into the dialogue by answering in particular ways and by guiding the types of questions asked. … Each researcher is willing to enter into this dialogue because each recognises the potential fallibility of their own view of reality and so the potential for modification. … [The] practice or the 'doing' of physical geography requires the investigator to continually shift from the 'objective' of the entity under study to the 'subjective' of the mental categories and concepts used in order to understand the entity. The continual shifting and informing of practice by the 'real' and the 'mental' means that neither can be understood in isolation. Practice brings to the fore the relational nature of understanding in physical geography and the haziness of seemingly concrete concepts such as entities.
>
> (Inkpen and Wilson, 2015, pp. 43–44)

Within human geography the 'spatial science' tradition is also frequently perceived as positivist. This field of research emerged in the 1960s with the aim of making geography more scientific. It involves the application of quantitative methods to identify spatial regularities in the distribution of social and economic phenomena and the development and testing of causal explanations and theories. Because it uses numerical data, develops generalisations and law-like statements and searches for explanations, spatial science is often labelled as positivist and is assumed to have the problems outlined earlier. Quantitative methods may also be considered to be positivist by some human geographers and therefore to be epistemologically unacceptable. Some have gone further and argued that spatial science and quantitative methods are also morally unacceptable, because of a perceived link between positivism, quantification and conservative politics and policies. Therefore, many geographers have rejected quantitative methods altogether, and teaching in these areas in universities has declined.

The portrayal of spatial science and quantitative methods as positivist has been vigorously contested. Those who regard themselves as spatial scientists and/or quantifiers have argued that their research is not, and never has been, strictly positivist, because they acknowledge that observation is theory-driven and its results require interpretation that is inevitably subjective. They also point out that developments in spatial science, and in the quantitative methods and mathematics it uses, have made it much closer to non-positivist research. Furthermore, they argue that progressive and critical research needs to include quantitative analysis, because this produces 'the kind of evidence that meets the standards of proof established by the state, corporations, or general public discourse' (Wyly, 2009, p. 318). Research demonstrating the extent of poverty, or morbidity, or social disadvantage, for example, requires quantitative methods if it is to convince policy makers. Linda Peake, a feminist geographer, acknowledges the value of quantitative methods to progressive and critical research. In commenting on the near universal avoidance of these methods by feminist geographers, she argues that:

> The desire by feminist geographers ... to refrain from using quantitative methods is not an innocent one. ... Decisions about whether to believe what feminists say depend to a large degree on judgments of our competence and expertise. The normalization of qualitative methods restricts the grounds on which such judgements can be made. It also affects what kinds of questions are seen as central and what are seen as less important within the discipline, thus narrowing the field and type of knowledge we produce, reducing the attribution of epistemic authority upon feminist geographers.
>
> (Peake, 2008, p. 263)

As Mei-Po Kwan and Tim Schwanen (2009, p. 283) argue, 'the antagonism between critical and quantitative geographies is not beneficial to the discipline', as quantification can make a major contribution to issues of concern to critical geographers. Eric Sheppard also argues that the quantitative-qualitative divide is a 'disciplinary cultural construct' that developed during the evolution of Anglophone geography and is artificial and unproductive. He writes

> Realizing the potential of geographical thinking will mean training coming generations to appreciate and perform quantitative, qualitative, and mixed methods. Some of this cross-training can now be found in 'land change science,' which is expanding beyond its analytical, quantitative, and science-led roots.[2]
>
> (Sheppard, 2022, p. 19)

Constructivism

Constructivism, also known as interpretivism, is a philosophy of knowledge that developed in opposition to positivism. It includes a wide range of individual approaches that differ in some of their beliefs and practices, and the description below may not apply in full to all of them.

- Constructivists are often portrayed as rejecting the belief that there is a real world. Some do, but many accept that there is a reality, while differing from positivism in denying that we can directly access it through observation. Consequently, reality is what we think it is and not something objective and independent of our thinking. Rob Moore (2014, p. 30) neatly describes this difference between positivism and constructivism by suggesting that for positivists, knowledge is produced by the real, whereas for constructivists, the real is produced by knowledge. David Marsh, Selen Ercan and Paul Furlong (2018, p. 183) add the point that 'We are not claiming that such researchers do not acknowledge that there are institutions and other social entities. Rather, they contend that this "reality" has no social role/causal power independent of the agent's/group's/society's understanding of it'.
- Constructivist geographers focus on understanding people's beliefs, perceptions and interpretations of the world, and how these influence how they act in and create the world. For example, the perceptions people have of different places in a city influence whether they visit them or not, which, in turn, influences what the places themselves are like. People's actions are influenced not by what these places are really like, but by their perceptions of what they are like.
- A constructivist view of knowledge is relativist, arguing that different societies, cultures and social groups have different ways of viewing the world and different ways of judging the validity of their knowledge. Because there are no criteria to decide between these different ways, and no directly accessible external reality against which to test our experience of the world, there can be no objective or universal knowledge truths. Relativism can also be applied to individuals, because knowledge is what people construct from their own experience of the world, and their different interpretations may be equally valid.
- Constructivists do not seek to develop generalisations, because they acknowledge people's different experiences and perceptions of the world. Furthermore, they do not attempt to predict human behaviour. This is because they do not consider that people are subject to laws that determine their actions, as humans can exercise agency and determine their own actions. However, some contemporary constructivists do attempt to explain how the world works, but mostly only at the scale of an event or a particular social phenomenon. At the same time, they recognise that how the world worked in this instance could have been different if people had made different choices.

- Some forms of constructivist thinking are a reaction to a perceived dominance of social science by male, white, middle-class academics and a justified belief that accepted disciplinary knowledge reflects their experience of the world. The development of feminist and post-colonial geographies, for example, recognised the different perceptions and experiences of women and Indigenous peoples.
- Following from the last point, constructivists (and many others) recognise that people's interpretations of what they observe can be influenced by their gender, class, race and other personal characteristics. Angela Saini, in her book on patriarchy, writes about a female graduate student in Near Eastern archaeology who:

  noticed [male] experts tying themselves in knots trying to explain why the bodies of women discovered in the ancient Sumerian city of Ur had been buried with royal objects or with weapons. At times [she explained], the integrity of hard archaeological data was being questioned because it didn't align with gendered expectations. Researchers seemed more willing to believe that their data might be wrong than to entertain the possibility that their assumptions might be.

  (Saini, 2023, pp. 141–142)

Criticisms of constructivist philosophies include:

- Its focus on people's perceptions and interpretations of why they act the way they do. This may lead to a neglect of the external structural forces, such as institutions or the economy, that influence and constrain their behaviour and of which people may be unaware.
- Its limited usefulness for policy. James Scotland (2012, p. 12) writes that 'Generalizations which are deemed useful to policy makers are often absent because [constructivist] research usually produces highly contextualized qualitative data, and interpretations of this data involve subjective individual constructions'.
- The lack of any objective ways to evaluate the validity of constructivist research outcomes. In response, constructivists agree that their research cannot be tested through the methods of positivism, which are often quantitative, but argue that there are ways to ensure that their qualitative research is rigorous. For example, Elaine Stratford and Matt Bradshaw argue that:

  Ensuring rigour in qualitative research means establishing the *trustworthiness* of our work. Research can be construed as a kind of 'hermeneutic [interpretive] circle' starting from our interpretive community [disciplinary colleagues], and involving our research participant community and ourselves, before returning to our interpretive community for assessment. This circle is a key part of ensuring rigour in qualitative research; our participant and interpretive communities check our work for credibility and good practice. In other words, trust in our work is not assumed but has to be earned.

(Stratford and Bradshaw, 2021, p. 102)

- The claim that there can be no scientific 'truths'. This seems to ignore the fact that much of our life benefits from the application of these 'truths' to technological, medical and other fields.

- The belief in relativism. Rob Moore argues that constructivism contains a logical contradiction, in that: 'if it is true that all truth is relative, then there must be one truth that is not, namely the truth that all truth is relative' (Moore, 2014, p.28). It also removes the need for the types of powerful thinking described in Chapter 2, because: 'Given that relativism allows people to construct their own 'personal truths', critical thinking turns out to be unnecessary .... [so] there is no need to evaluate ideas or search for alternatives, because all ideas are equally trustworthy and justifiable' (Hyytinen et al., 2014, p. 4).

Relativism can also undermine scientific arguments on issues that are opposed by powerful interests. As Elvin Wyly (2011, p. 894) writes: 'If science is made, [so] conservatives understand, then it can be made up: what began on the left as a sensitivity to the social construction of knowledge became, on the right, a coordinated program of creating well-financed lies'. An example of this point is conservative rejection of the evidence for the effects of fossil fuel use on climate change, using the argument that scientific opinions are only opinions, and therefore alternative opinions are equally valid.

However, many thinkers who might be regarded as having a constructivist philosophy disagree that their approach is relativist in the sense that any opinion goes. They argue that the methods they use to test their ideas and conclusions, such as the ones identified by Stratford and Bradshaw earlier, ensure that they are robust. They might also add that their approach is more 'objective' than positivism because it explicitly recognises the influence of the researcher's own social position (Zipin et al., 2015).

Despite these criticisms, constructivist approaches to gaining knowledge are a strong part of contemporary human geography, and for teachers they are also the basis of constructivist methods of teaching, as will be noted later.

Realism

Realism is sometimes described as a philosophy of knowledge that is in-between positivism and constructivism, and it also has a number of variations. This section focuses on critical realism, which has had some influence on research in both physical and human geography.

Critical realists agree with positivists that there is a reality, both natural and social, that is independent of our knowledge of it, but like constructivists recognise that what we perceive of that reality is a human construct and imperfect and subject to revision. They argue that while we cannot directly observe much of reality, especially in the social world, we can attempt to understand it through events that can be observed. Priscilla Alderson explains that critical realists consequently understand knowledge and reality at three levels:

> The empirical level is our sensed experiences, perceptions and interpretations, our thinking and talking. The actual level includes actual activities, events, people, objects, relations and structures, the being and doing of the independent real world, an infinite reality that we can know only slightly and fallibly. At the third level, the real, are powerful causal mechanisms usually only seen in their effects. Natural causal mechanisms unseen by the naked eye include gravity, microbes, molecular and genetic structures. Social causal mechanisms include human values and motives, justice and inequalities of class and race.
>
> (Alderson, 2020, p. 97)

For example, a concept such as patriarchy cannot be directly observed, but its existence and influence can be theorised from the observable outcomes that it produces, as in the example from archaeology mentioned earlier.

While agreeing with constructivists that what we perceive of the world is a human construct and that our observations and interpretations cannot be totally objective, critical realists differ from them in rejecting relativism. They argue that some knowledge statements are better than others because they have been tested against our experience of a real world and found to 'work'. They have also been reviewed and critiqued by others and have not yet been rejected. Their acceptance, however, is always provisional.

Critical realists also aim to try to explain the world, not just understand it like most constructivists, but also have a different view of causation to positivists. They recognise that the way social reality is constructed by people influences how they, as individuals or organisations, behave and act, and therefore produce the outcomes that we can observe. Consequently, the meanings and understandings that people have about their world and themselves can be causal factors. As Marsh, Ercan and Furlong write, in relation to research in political science

> Modern critical realism acknowledges two points. First, while social phenomena exist independently of our interpretation of them, our interpretation and understanding of them affects outcomes. So, structures do not determine; rather, they constrain and facilitate. Social science involves the study of reflexive agents who interpret and change structures. Second, our knowledge of the world is fallible; it is theory-laden. We need to identify

> and understand both the external 'reality' and the social construction of that 'reality' if we are to explain the relationship between social phenomena.
>
> (Marsh et al., 2018, p. 194)

Unlike positivists, critical realists contend that causal relationships cannot be directly observed but instead only inferred as a way to explain observations. Rather than looking for regularities in which one event is always accompanied by another, and identifying this as a causal relationship, as in positivism, critical realism tries to uncover the mechanisms that produce these events and the contexts in which they occur. Critical realists contend that the same causal mechanism may produce different outcomes in different contexts, and Joseph Holden explains the value of this for physical geography, and by implication for all geography:

> This approach brings together the idea of general laws and local events. It helps us understand how local factors influence the mechanisms to produce an individual example. In this way it is different from traditional positivist approaches which are just interested in generalizations and iron out any irregular or unusual cases. In realism the case study itself becomes of interest in helping us to understand the world around us. This is because what causes change is the interaction of mechanisms with particular places. This is particularly useful for geography, which has a tradition of examining case studies.
>
> (Holden, 2017, p. 13)

For social phenomena, critical realist geographers attempt to identify the underlying structures that are producing the outcomes they observe. Examples of these structures are capitalism and patriarchy, which influence or constrain how people and organisations behave in ways they may not be aware of. In such cases, the explanations that people give for their behaviour may ignore the 'deep' causes of that behaviour, as was noted earlier as a criticism of constructivism. Marxism is the classical form of critical realism and has been influential in human geography. Political ecology, also influential in geography, is another form, as it examines how social, economic and political structures and inequalities affect environmental outcomes, such as land degradation. Some geographers have emphasised these structural aspects of critical realism, while others have stressed the significance of context in explaining events, as in the quotation from Holden earlier. Andrew Sayer (2000, pp. 23–25), for example, explains how firms in the same industry in England performed differently under the same economic pressures, because the context in which each firm operated was different.

Critical realism therefore has an ontology similar to positivism and an epistemology similar to constructivism. Consequently, it can be argued that it avoids the empirical narrowness of positivism and the relativism of some forms of constructivism, and this has been an attraction for some geographers. Critical realism is also compatible with a variety of research methods, including both quantitative and qualitative techniques. Social realism is a version of critical realism and is the philosophical foundation of the concept of powerful knowledge, which has had a significant influence on school geography and will be briefly discussed in the next chapter.

Positivism, constructivism and critical realism compared

Marsh, Ercan and Furlong, writing from the perspective of political science, have a good summary of the major differences between the three philosophies, which might help readers to decide which one they follow in practice:

> So, a positivist looks for causal relationships, tends to prefer quantitative analysis and wants to produce 'objective' and generalisable findings. A researcher from within the interpretivist [constructivist] tradition is concerned with understanding, not with predictive explanation, focuses on the meaning that actions have for agents, tends to use qualitative evidence and offers his/her results as one interpretation of the relationship between the social phenomena studied. Critical realism is less easy to classify in this way. The realists are looking for causal relationships, but argue that many important relationships between social phenomena can't be observed. This means that they may use quantitative and qualitative data. The quantitative data will only be appropriate for those relationships that are directly observable. In contrast, the unobservable relationships can only be established indirectly; we can observe other relationships which, our theory tells us, are the result of those unobservable pre-relationships.
>
> (Marsh et al., 2018, p. 185)

Each approach has its strengths, and each provides a different view of the world. Positivist geographers (most of whom are not strictly positivist) can describe and analyse issues at a scale that can be used in policy and can identify possible causal relationships. However, while their quantitative methods may be able to identify the likely causal processes involved in these relationships, they cannot explain how they work. Constructivist geographers can help us understand the diversity of people's experiences of the world and uncover the meanings that influence human behaviour that may be missing from positivist research. They can also uncover and highlight the lives of less powerful and often marginalised social groups. Finally, critical realist geographers can provide deeper explanations than positivist approaches and can challenge conventional power structures and thinking. Given the complementary strengths of the three philosophies of knowledge, it may be tempting to want to combine them, but the answer to whether this is possible or even desirable seems, for most people, to be no, because they make different assumptions about what we can know, and how we can know it. On the other hand, each approach can add to a fuller understanding of often complex problems, and each should be respected by those from different schools of thought.

What does all this mean for geographical education?

If you are a geography teacher in a school, or preparing to become one, you may be asking whether any of this discussion of competing philosophies of knowledge has any relevance to your work. It has, because it influences your view of your role as a teacher, what you think your students should learn and how they should learn it. This section of the chapter briefly discusses two ways in which the philosophies might be relevant to teaching in schools.

Personal epistemologies

A way of connecting types of epistemologies with teaching is through the findings of research into the personal epistemologies of children and adults. One body of research describes three stages in the development of epistemic[3] beliefs, which are people's beliefs about how knowledge is constructed and validated, i.e., their personal epistemologies (Brownlee et al., 2014; Grossnickle et al., 2017).

- Absolutist beliefs view knowledge as facts which are absolute, and if there are competing facts, only one can be correct. Individuals who hold absolutist beliefs are likely to think that learning is about memorising and repeating the information provided by the teacher. Similarly, teachers with absolutist beliefs are likely to view children's learning as passive and not requiring critical thinking.
- Multiplist beliefs view knowledge as constructions based on one's personal opinions. Competing opinions have equal value, so individuals who hold such beliefs are likely to listen to and value the opinions of others in a class, but not evaluate them. Teachers with multiplist beliefs are also unlikely to encourage critical thinking about the validity of knowledge opinions and instead promote an acceptance of some ideas that may lack substance.
- Evaluativist beliefs view knowledge as judgements based on an evaluation of a variety of types of evidence. If there are competing judgements, one can be selected as the best through a comparative analysis of the evidence and the supporting arguments. However, judgements are provisional and open to challenge from new evidence and new ideas. Individuals who hold evaluativist beliefs are likely to construct knowledge through a review of multiple perspectives. A teacher with evaluativist perspectives is likely to view children's learning as one in which they construct knowledge, think in complex ways about ideas and engage in argument to verify and justify their

judgements. This is the best context for the development of thinking skills, and the approach advocated in this book.

These stages describe a progression in which individuals:

> develop in sophistication from a complete trust in authority and use of authority as a justification for knowledge to complete distrust in authority and skepticism of any justifications for knowledge, and eventually to more moderate beliefs that evidence must be carefully considered and weighed.
> (Grossnickle et al., 2017, p. 4)

The three-stage developmental progression described above is rigid and needs qualification. One outcome of research into people's epistemic beliefs is that the types of evidence and the ways in which they are evaluated differ between disciplines, and consequently teachers should show students the particular ways of individual disciplines, and not confine teaching to generic methods. History, for example, has well-developed methods of evaluating claims about past events. Geography, on the other hand, has very wide-ranging types of knowledge, some of which can be evaluated using scientific reasoning, while others require methods similar to historical enquiry and those used in other humanities subjects. Another lesson from the research is that epistemic competence, or the ability to assess evidence and evaluate knowledge claims, does not happen by chance. This is because 'learning environments that promote the development of epistemic competence are [ones] ... where evidence is weighed and critiqued, and where discussions about the nature of evidence common to domains and disciplinary communities are part of the classroom discourse' (Grossnickle et al., 2017, p. 12).

This is another important role for teachers. A suggested technique to develop epistemic competence is to present students with problems where there are conflicting claims, a variety of types of evidence, no obvious best answer, and therefore scope for students to evaluate, reason and debate. This is where thinking skills can be developed, through their application to a specific problem. Chapter 9 discusses this and other methods for developing thinking in some detail.

Constructivism as a theory of teaching

Another example of the relevance of philosophies of knowledge to teaching is that, as many teachers will be aware, constructivism has informed a theory of teaching in which students are helped to construct their own knowledge. In geography, constructivism is the epistemological foundation for inquiry-based learning, a popular method until examination pressures intervene. However, if you reject constructivism and its claim that all knowledge has equal validity, then you have the problem of ensuring that students avoid 'incorrect' constructions of knowledge. As Michael Ford (2010, p. 265) observes: 'Many teachers note a tension in constructivist teaching, a tension between allowing students to construct their own sense of disciplinary ideas and ensuring that the sense they make is correct'. He suggests that a way to overcome this problem is to teach students to evaluate their own knowledge, using the methods that disciplines have developed to judge the validity of knowledge constructions. Stemhagen, Reich and Muth have a similar answer when they argue that:

> a primary challenge to constructivist educators is to find ways for *students* to become skilled, not only at constructing knowledge, but also at evaluating it—judging its worth—in disciplined ways. It is our contention that the inclusion of judgment is the fulfillment of the promise of empowerment and meaningfulness that is implicit in learner-centered constructivist pedagogy.
>
> (Stemhagen et al., 2013, pp. 58–59)

Indigenous knowledges

Indigenous societies provide another body of knowledge, through place-based ways of knowing. In recent years, Indigenous knowledges have become environmentally, politically and educationally significant in a number of countries, and especially in Canada, the USA, Australia, New Zealand and South Africa. These are countries of European colonisation where the knowledge of the Indigenous peoples has either survived or been revived, but traditional environmental knowledge has also survived in parts of Europe. It is a body of knowledge that has both similarities to and differences from the knowledge produced by the approaches discussed earlier in this chapter. This short section discusses some of its characteristics.

Indigenous knowledge is local, the product of and embedded in a specific cultural group. There are consequently many Indigenous knowledges, not just one, and the knowledge of one group is not necessarily the same as in neighbouring groups. In this sense, Indigenous knowledge is context-dependent, although the way of thinking is likely to be transferable to other places. Cameron Muir, Deborah Rose and Phillip Sullivan write:

> [Australian] Aboriginal people overtly acknowledge the agency of place in everyday interactions and its role in constituting and maintaining knowledge ...[and] the importance of place, of the situated context of knowledge, is deeply embedded in Aboriginal people's language and in ways of speaking in and about country.
>
> (Muir et al., 2010, pp. 260–261)

While this viewpoint may differ from Western science's assertion of the universality of its knowledge, it is similar to the geographer's contention that places are unique and that their uniqueness influences the outcomes of environmental and social processes, as will be discussed in Chapter 4.

Because of this localness, it has been argued that Indigenous knowledges may not be able to deal with larger-scale changes such as climate change. On the other hand, Fikret Berkes describes how Inuit people in Canada, in collaboration with scientists, have built a knowledge base of climate change data because of their 'way of observing, discussing and making sense of new information', which he describes as Indigenous ways of knowing (Berkes, 2009, p. 153). It therefore may be useful to distinguish Indigenous knowledges from the processes used to gain that knowledge.

Indigenous environmental knowledge is highly empirical, based on careful observations over many generations, but the ways Indigenous people perceive the world may be different to Western science, although no less empirical. An example comes from a research project that asked Indigenous and non-Indigenous fishing experts to classify fish into groups, without specifying the types of groups. The research reported that: 'Non-Native experts tend to sort taxonomically ("these fish belong to the bass family") while Menominee experts are more likely to sort ecologically ("these fish live in cool, fast moving waters")' (Bang, 2018, p. 153).

The justification for environmental knowledge may be spiritual rather than empirical. For example, in Central Australia, there are several totemic sites where hunting of the red kangaroo is banned by Aboriginal law. Ecological studies show that these sites coincide with the most favourable habitat for the species, so are the best sites to select to conserve the red kangaroo and were presumably chosen for this reason by Aboriginal people long ago. This is a conservation method identical to contemporary Marine Protected Areas in which fishing is banned to conserve fish stocks. However, Aboriginal people do not explain the sites where hunting is banned as a conservation measure but as created by the actions of the Ancestral Beings.

Indigenous knowledges are about more than the practical environmental knowledge needed for communities to survive. While much of it is about weather and seasons (Box 1.1), plants and animals, food production, water resources, the use of fire, tool-making, medicines and healing, some is also about less day-to-day matters such as law, time, astronomy and spirituality.

Box 1.1 Australian Aboriginal seasonal calendars

Aboriginal seasonal calendars are ecological calendars and different to the four-seasons calendar familiar to mid-latitude Europeans because they are based on changes in the environment, such as the flowering of plants, the behaviour of animals and the availability of food and water, and not on fixed dates. They tell people what resources are available and where, when to move to use them, when to burn to maintain the landscape and when to hold major ceremonies that require large quantities of food. Rose provides an example from the Yarralin People in the Northern Territory:

> When the flowers of the jangarla tree (Sesbania formosa) fall into the water, the barramundi start to bite. When the green flies arrive, a certain species of bush plum is ripe. When the cicadas start singing, the turtles begin to gain fat. When the brolga returns, the dark catfish become active: the river will start to flow soon.
>
> (Rose, 1997, p. 16)

These calendars are now recognised as a major store of environmental knowledge that has been gathered over thousands of years by people whose lives completely depended on its accuracy and comprehensiveness. David Jones and Philip Clarke describe them as:

> unwritten management plans in their own right that can have direct applicability into Western land management regimes. … Comprising 'Indigenous biocultural knowledge' in the ecosystem science and management discipline, these seasonal models incorporate plant, aquatic and terrestrial animals, fire, climate, drought, flood, and astronomical information (that could be termed a 'spectrum of knowledge') drawn from Aboriginal sources ….
>
> (Jones and Clarke, 2018, p. 52)

In Indigenous knowledges, everything is alive, including rocks, stars and sand dunes, and human and non-human life are all interconnected in reciprocal relationships. Indigenous peoples consequently have a holistic view of the world, the opposite of the reductionist approach of science. Berkes writes: 'Indigenous knowledge systems seem to build holistic pictures of the environment by considering a large number of variables qualitatively, while science tends to concentrate on a small number of variables quantitatively' (Berkes, 2009, p. 154). Western geography can learn from this, as interconnection is a key concept in geographical thinking and is discussed in Chapter 3.

Indigenous knowledge is frequently and incorrectly portrayed as inherited from the past, and fixed and static. Berkes, on the other hand, contends that it 'evolves all the time and involves constant learning-by-doing, experimenting and knowledge-building' (Berkes, 2009, p. 154). Many Indigenous people have adapted to the environmental changes produced by European colonisation, such as the spread of feral animals and introduced weeds. Many have adopted aspects of Western knowledge that are useful to them. Jennifer Carter and David Hollinsworth (2017) argue for the importance of recognising the 'intercultural' nature of the world of many Indigenous people, who are adapting, changing and blending Indigenous and Western knowledge and ways of thinking.

A useful summary description of Indigenous knowledges has been developed by The Intergovernmental Science-Policy Platform on Biodiversity and Ecosystem Services (IPBES):

> Indigenous and local knowledge systems are in general understood to be bodies of integrated, holistic, social and ecological knowledge practices and beliefs pertaining to the relationship of living beings, including people, with one another and with their environments. Indigenous and local knowledge is grounded in territory, is highly diverse and is continuously evolving through the interaction of experiences, innovations and various types of knowledge (written, oral, visual, tacit, gendered, practical and scientific). Such knowledge can provide information, methods, theory and practice for sustainable ecosystem management. Many Indigenous and local knowledge systems are empirically tested, applied, contested and validated through different means in different contexts.
>
> (Quoted in Woodward et al., 2020, p. 6)

These brief statements about Indigenous knowledges suggest that they are similar to some of the geographical thinking discussed in this book but often go well beyond it. Non-Indigenous geographers could learn from and include these diverse perspectives in their teaching, particularly in relation to interconnectedness, holism and the significance of place. They also teach students that there can be more than one way of perceiving and relating to the world.

Conclusion

This chapter has described four different answers to the questions 'what can we know' and 'how can we know it'. Which one or ones you identify with will influence what you teach and how you teach it. The key questions to ask yourself are:

- Is there a reality that is independent of my perception of it?
- How objective can I be in my knowledge of the world?
- How can I explain what I observe?
- How do my answers to these questions influence what and how I teach?

Summary

- Ontology is about the nature of reality—about whether there is a real world out there, independent of what we think we know about it or is the world what we construct it to be.
- Epistemology is about what we can know about this world, how we can find this knowledge and how we can validate it.
- Positivism views the world as real, independent of our perception of it and able to be known through our observations.
- Positivism attempts to explain the patterns and regularities found through observation by constructing generalisations and identifying causal relationships.
- In positivism, hypotheses about causal relationships must be possible to disprove.
- Despite its claims, positivism is not fully objective.
- Physical geography and spatial science are not strictly positivist.
- Quantification need not be positivist and can contribute to critical geography.
- Constructivism argues that reality is a human construction and not independent of what we think about it.
- Constructivists seek to understand, not explain, and do not try to develop generalisations.
- A constructivist view of knowledge is relativist, although some versions have methods for evaluating the validity of knowledge claims.
- Constructivism is criticised for its neglect of the external structural forces that influence human behaviour, its limited usefulness for policy and its relativism.
- Realists agree that there is a reality that is independent of our knowledge of it but recognise that what we perceive of that reality is a human construct, and imperfect and subject to revision.
- Realists reject relativism.
- Realists look for the mechanisms that produce observable events. They recognise that the same causal mechanism may produce different outcomes in different contexts, which for geography are places.
- For social phenomena, critical realist geographers attempt to identify the underlying structures that are producing the outcomes they observe.
- An understanding of positivism, constructionism and realism will help teachers to explore the personal epistemologies of their students and evaluate the use of constructivist methods of teaching.
- Indigenous knowledges have both similarities to and differences from the three others discussed in the chapter and much to contribute to Western geographical thinking.

How could you use this chapter in teaching?

- Perhaps the most important question to regularly ask students is: how do we know? What is the reasoning and evidence used to support a statement about some piece of knowledge? Is it possible to say that some knowledge claims are better than others? When is a statement about knowledge more than an opinion?
- Another useful question is exploring why people have different opinions about something, such as climate change, sustainability, safe and unsafe places or spatial inequality.
- In a class discussion, students could be asked to think about whose experience is missing from the debate that might represent a different opinion. For example, in countries such as Australia, where the majority of the population is of European origin, many students have had no personal experience of racism and may argue that it does not exist. Aboriginal Australians, and those with Asian, African or Middle Eastern ancestry, are likely to have a rather different opinion. Experience and opinions also may differ amongst students according to gender, physical ability, social background or family income.
- In countries where Indigenous knowledge is still significant, teachers could use local examples to show that there are different ways of viewing the world.

Notes

1. These categories have been borrowed and adapted from Johnston (1990); Marsh et al. (2018); Grix (2004); and Firth (2013).
2. See Chapter 10.
3. Epistemic is derived from the Greek term episteme, which means knowledge, what is known, or the way of knowing. This term is typically used as an adjective, implying 'of or relating to knowledge' (Kitchener, 2011, p. 92).

Useful readings

Firth, R. (2013). What constitutes knowledge in geography? In D. Lambert & M. Jones (Eds.), *Debates in geography education* (pp. 59–74). Routledge.
An accessible introduction to ideas about knowledge and their implications for the school curriculum.

Kitchen, R. (2006). Positivistic geographies and spatial science. In S. Aitken & G. Valentine (Eds.), *Approaches to human geography* (pp. 20–29). Sage.

Sayer, A. (2006). Realism as a basis for knowing the world. In S. Aitken & G. Valentine (Eds.), *Approaches to human geography* (pp. 98–106). Sage.

A short and readable explanation of realism, and how it differs from positivism and constructivism, by one of its leading exponents.

Moore, R. (2014). Social realism and the problem of the problem of knowledge in the sociology of education. In B. Barrett & E. Rata (Eds.), *Knowledge and the future of the curriculum* (pp. 23–40). Palgrave Macmillan.

A more difficult but deeply thoughtful discussion of positivism, constructivism and realism, with an emphasis on their relevance to education.

Couper, P. (2015). *A student's introduction to geographical thought: Theories, philosophies, methodologies*. Sage.

For those wanting more depth, this book is by a physical geographer and is illustrated with examples from both physical and human geography. This is a major strength, as other books on geographical thought only examine one side of the subject. A theme running through the book is how different approaches to geography, from positivism to complexity theory, might examine a beach. Written for undergraduate students, and readable.

References

Alderson, P. (2020). Powerful knowledge, myth or reality? Four necessary conditions if knowledge is to be associated with power and social justice. *London Review of Education*, *18*(1), 96–106. https://doi.org/10.18546/LRE.18.1.07

Bang, M., Marin, A., & Medin, D. (2018). If Indigenous Peoples stand with the sciences, will scientists stand with us? *Daedalus*, *147*, 148–159. https://doi.org/10.1162/DAED_a_00498

Berkes, F. (2009). Indigenous ways of knowing and the study of environmental change. *Journal of the Royal Society of New Zealand*, *39*(4), 151–156. https://doi.org/10.1080/03014220909510568

Brownlee, J., Curtis, E., Chesters, S. D., Cobb-Moore, C., Spooner-Lane, R., Whiteford, C., & Tait, G. (2014). Pre-service teachers' epistemic perspectives about philosophy in the classroom: It is not a bunch of 'hippie stuff'. *Teachers and Teaching*, *20*(2), 170–188. https://doi.org/10.1080/13540602.2013.848565

Carter, J., & Hollinsworth, D. (2017). Teaching Indigenous geography in a neo-colonial world. *Journal of Geography in Higher Education*, *41*(2), 182–197. https://doi.org/10.1080/03098265.2017.1290591

Firth, R. (2011). Debates about knowledge and the curriculum: Some implications for geography education. In G. Butt (Ed.), *Geography, education and the future* (pp. 141–164). Continuum.

Ford, M. J. (2010). Critique in academic disciplines and active learning of academic content. *Cambridge Journal of Education*, *40*(3), 265–280. https://doi.org/10.1080/0305764X.2010.502885

Grix, J. (2019). *The foundations of research* (3rd ed.). Red Globe Press.

Grossnickle, E. M., Alexander, P. A., & List, A. (2017). The argument for epistemic competence. In A. Anschütz, H. Gruber, & B. Moschner (Eds.), *Wissen und lernen: Wie epistemische Überzeugungen Schule, Universität und Arbeitswelt beeinflussen* [Knowledge and learning in the perspective of learners and instructors: How epistemic beliefs influence school, university and the workplace]. Waxmann-Verlag. Available from: https://www.researchgate.net.

Hanson, S. (2004). Who are 'we'? An important question for geography's future. *Annals of the Association of American Geographers, 94*, 715–722. https://doi.org/10.1111/j.1467-8306.2004.00425.x
Holden, J. (2017). *An introduction to physical geography and the environment* (4th ed.). Pearson.
Hyytinen, H., Holma, K., Toom, A., Shavelson, R. J., & Lindblom-Ylanne, S. (2014). The complex relationship between students' critical thinking and epistemological beliefs in the context of problem solving. *Frontline Learning Research, 2*(5), 1–25. https://doi.org/10.14786/flr.v2i4.124
Inkpen, R., & Wilson, G. (2013). *Science, philosophy and physical geography* (2nd ed.). Routledge.
Johnston, R. J. (1990). Viewpoint: Exploring the role of geography. *Geography Research Forum, 10*, 91–103. https://grf.bgu.ac.il/index.php/GRF/article/view/82
Jones, D. S., & Clarke, P. A. (2018). Aboriginal culture and food-landscape relationships in Australia. In J. Zeunert & T. Waterman (Eds.), *Routledge handbook of landscape and food* (pp. 41–60). Routledge.
Kwan, M.-P., & Schwanen, T. (2009). Quantitative revolution 2: The critical (re) turn. *The Professional Geographer, 61*(3), 283–291. https://doi.org/10.1080/00330120902931903
Marsh, D., Ercan, S. A., & Furlong, P. (2018). A skin not a sweater: Ontology and epistemology in political science. In V. Lowndes, D. Marsh, & G. Stoker (Eds.), *Theory and methods in political science* (4th ed., pp. 177–198). Palgrave.
Muir, C., Rose, D., & Sullivan, P. (2010). From the other side of the knowledge frontier: Indigenous knowledge, social–ecological relationships and new perspectives. *The Rangeland Journal, 32*, 259–265. https://www.publish.csiro.au/RJ/RJ10014
Peake, L. J. (2015). The Suzanne Mackenzie Memorial Lecture: Rethinking the politics of feminist knowledge production in Anglo-American geography. *The Canadian Geographer* 59(3), 257–266. https://doi.org/10.1111/cag.12174
Rose, D. (1997). When the rainbow walks. In E. K. Webb (Ed.), *Windows on meteorology: Australian perspective* (pp. 12–17). CSIRO Publishing.
Saini, A. (2023). *The patriarchs: How men came to rule.* 4th Estate.
Sayer, A. (2000). *Realism and social science.* Sage.
Scotland, J. (2012). Exploring the philosophical underpinnings of research: Relating ontology and epistemology to the methodology and methods of the scientific, interpretive, and critical realist research paradigms. *English Language Teaching, 5*(9), 9–16. https://doi.org/10.5539/elt.v5n9p9
Sheppard, E. (2022). Geography and the present conjuncture. *EPF: Philosophy, Theory, Models, Methods and Practices, 1*(1), 14–25. https://doi.org/10.1177/26349825221082164
Stemhagen, K., Reich, G. A., & Muth, W. (2013). Disciplined judgment: Toward a reasonably constrained constructivism. *Journal of Curriculum and Pedagogy, 10*, 55–72. https://doi.org/10.1080/15505170.2012.724360
Stratford, E., & Bradshaw, M. (2005). Rigorous and trustworthy: Qualitative research design. In I. Hay & M. Cope (Eds.), *Qualitative research methods in human geography* (5th ed., pp. 92–106). Oxford University Press.
Woodward, E., Hill, R., Harkness, P., & Archer, R. (Eds.) (2020). *Our knowledge our way in caring for country: Indigenous-led approaches to strengthening and sharing our knowledge for land and sea management. Best Practice Guidelines from Australian*

experiences. NAILSMA and CSIRO. https://www.csiro.au/-/media/LWF/Files/OKOW/OKOW-Guidelines_FULL.pdf

Wyly, E. (2009). Strategic positivism. *The Professional Geographer*, *61*(3), 310–322. https://doi.org/10.1080/00330120902931952

Wyly, E. (2011). Positively radical. *International Journal of Urban and Regional Research*, *35*(5), 889–912. https://doi.org/10.1111/j.1468-2427.2011.01047.x

Zipin, L., Fataar, A., & Brennan, M. (2015). Can social realism do social justice? Debating the warrants for curriculum knowledge selection. *Education as Change*, *19*(2), 9–36. https://doi.org/10.1080/16823206.2015.1085610

Chapter 2

Thinking geographically

What does it mean to think, and to think geographically? Thinking is a cognitive process, and its different forms are described by verbs such as analysing and explaining. There are many forms of thinking, so which ones should be emphasised in school geography? This chapter discusses a way to answer this question, by using the concept of powerful knowledge to identify types of thinking. It then explains how these can be made geographical by applying the ways of thinking embodied in the discipline's major concepts.

Thinking

The concept of powerful knowledge provides a way of identifying some educationally valuable types of thinking. Powerful knowledge is an idea that was introduced into educational debates well over a decade ago by Michael Young, a British sociologist of education. It has been picked up by geography educators in several countries, adopted by the international GeoCapabilities Project (Lambert et al., 2015) and applied in several recent books for geography teachers.[1] Young argues that, while knowledge is a human construct and is fallible and never absolute, when produced within disciplinary communities and subjected to disciplinary critique, it is the best knowledge available. It is this that makes it powerful, and it this knowledge that should be taught in schools.

However, critics have argued that 'knowledge alone cannot exercise power', because this depends partly on the agency of the knower (Alderson, 2020, p. 98). In other articles (Maude, 2016, 2017), I have suggested that the word 'power' implies an ability or capacity to do something that has an effect or outcome, so powerful knowledge is knowledge that has powerful outcomes for those who are able to exercise it. Young in fact also describes knowledge as powerful because of 'what the knowledge can do or what intellectual power it gives to those who have access to it' (Young, 2008, p. 14). The passages below provide further insights into this interpretation of powerful knowledge:

DOI: 10.4324/9781003376668-3

> Knowledge is 'powerful' if it predicts, if it explains, if it enables people to envisage alternatives, if it helps people to think in new ways.
>
> (Young, 2015, n.p.)

> 'Powerful knowledge' is powerful because it provides the best understanding of the natural and social worlds that we have and helps us go beyond our individual experiences.
>
> (Young, 2013, p. 196)

> [Knowledge is powerful when] its concepts ... can be the basis for generalisations and thinking beyond particular contexts or cases.
>
> (Young, 2015, n.p.)

From these and other statements by Young, we can identify several types of knowledge that are powerful because of what they enable students to do, and those selected as relevant to this book are the ability to:

- Think about the world in new ways.
- Better understand and explain the natural and social worlds.
- Evaluate claims about knowledge, including one's own.
- Develop generalisations and apply them to new contexts.
- Forecast and evaluate alternative futures.[2]

Each of these abilities employs particular types of thinking, as shown in Table 2.1. Each also depends not only on knowledge but on how that knowledge can be applied. Thinking about the world in new ways, for example, gives students new questions to ask, and these may lead to better understanding and to knowledge that can be applied to solve new problems.

Geographical thinking[3]

The types of thinking in the table are common to many school subjects, so how can they be made 'geographical'? This can be achieved by interpreting them through geography's key concepts, because these identify the subject's distinctive ways of thinking. However, as there is no definitive list of what these concepts are, how do we decide which ones to choose? Eleanor Rawling pioneered thinking about this in English-language education in her 2007 guide for teachers on *Planning your Key Stage 3 geography curriculum*. In it she wrote that the big key concepts in geography 'may be thought of as standing at the top of a hierarchy of ideas [and] are the most abstract and generalised' (Rawling, 2007, pp. 23–24). Clare Brooks calls them hierarchical concepts, which are used 'to group the contents of the subject' and 'to

Table 2.1 Powerful abilities and powerful types of thinking

Powerful abilities	*Powerful types of thinking*
Think about the world in new ways	Geographical ways of perceiving the world
	Questioning
Better understand and explain the natural and social worlds	Organising and analysing
	Comparing
	Explaining and understanding
Evaluate claims about knowledge, including one's own	Evaluating
	Reflecting
Develop generalisations and apply them to new contexts	Generalising
	Applying
Forecast and evaluate alternative futures	Forecasting
	Evaluating
	Responding

represent ideas, generalisations or theories' (Brooks, 2018, p. 105). David Lambert describes them as

> large, organising ideas [that] underlie a geographical way of investigating and understanding the world. They are high level ideas that can be applied across the subject to identify a question, guide an investigation, suggest an explanation or assist decision making.
>
> (Lambert, 2017, p. 26)

Janis Fögele writes that they demonstrate 'consistent structures in the phenomena dealt with' (Fögele, 2017, p. 61). These ideas can be developed into the following criteria for a key geographical concept.

- The concept appears in most lists of geography's major concepts in both the academic literature and school curricula, so has been widely selected by practising geographers.
- It is at the top of a hierarchy of concepts of increasing complexity and abstractness. It can be thought of as 'key' because it synthesises and incorporates simpler and less abstract concepts and cannot be subsumed by an even bigger and more abstract one.[4]
- It can be applied to a great variety of topics and across different fields of the subject, so is 'key' in that it is widely used and gives geography a degree of unity and coherence.
- It has a number of functions, such as identifying topics worth studying and questions to ask, organising information, suggesting methods of analysis, forming generalisations, identifying possible explanations and providing a basis for public policies.

Place, space, environment and interconnection are the four concepts that fully meet these criteria, although interconnection may be called interdependence or interaction in some sources. They are central to geographical thinking and are the subject's biggest ideas. They give the subject coherence, linking the different topics studied in school geography through shared concepts and the ways of thinking they produce, as explained by the UK Geographical Association:

> Geography is a content-rich subject and concepts provide an underlying structure. Many topics in geography exemplify the same conceptual understanding, so it is important for learners to understand concepts so that they do not see geography as an accumulation of 'content' and 'facts'.
>
> (Geographical Association, n.d.)

Although the four concepts are also employed in other disciplines, such as ecology, archaeology, epidemiology, economics and sociology, in none are they as central to thinking and practice as in geography, and in none are they as frequently used in combination. Furthermore, when used in geography they may not be identical to their use in another discipline. Space, for example, is sometimes used in other social sciences to mean the same as the way place is described in this book. As used in geography, they are what makes the subject 'geographical', and they enable students to 'distinguish what is distinctively different about learning geography from other school subject areas' (Brooks, 2018, p. 106).

Other major geographical concepts that might have been selected are processes, systems, time or change, scale, region, nature, landscape and sustainability. Several of these do not qualify as 'key' because they are not at the top of a hierarchy of concepts. Processes and systems, for example, are subsidiary concepts of interconnection (as will be explained in Chapter 3), region is a subsidiary concept of place (because it is a type of place) and nature and landscape are subsidiary concepts of environment. This leaves time or change, scale and sustainability. These are important concepts, but they differ from the four key ones in the more limited range of their functions. Scale, for example, is largely an analytical concept, because it is mostly used in geography to analyse relationships by investigating them at different scales, or across scales. Time is also an analytical concept, because it can be used to explain phenomena by understanding how they have developed or changed over time. Sustainability, on the other hand, is largely an evaluative concept, because it is mostly used to assess the implications of an environmental change or the economic or demographic viability of a place. Human wellbeing is proposed as a second evaluative concept. Although it is rarely included as one of the disciplines' concepts, geographers implicitly use it, or the related but narrower concept of social justice, to assess the effects of environmental, economic and social change on people. It also explicitly adds an ethical or values dimension to geography, which writers such as David Mitchell and Alexis Stones have been arguing is missing from the discourse on powerful knowledge (Mitchell and Stones, 2022) and can be a stimulating topic for class debates. Sustainability and human wellbeing are of course closely connected, because human wellbeing (and even survival) depends on a sustainable environment. Scale, time, sustainability and human wellbeing have therefore been selected as core concepts in geography but do not meet the criteria for key concept status. In this book, place, space, environment, interconnection, scale, change, sustainability and human wellbeing are all termed core concepts, but only the first four are also *key* concepts.[5]

The key concepts are complex and very abstract ones and unlikely to make much initial sense to students. To understand them it is first essential to recognise that they are ideas that we think with, not objects that we study. For example, while places are parts of the Earth's surface that have been defined, named and given meaning by people, the concept of place is about ways of thinking that are based on the significance and influence of places. Second, they are not substantive concepts like 'city' or 'climate', which are about the substance of geography, but are meta-concepts, or concepts about concepts. Consequently, they are difficult to define in a single sentence because they have more than one dimension. As 'complex assemblages of interconnected smaller ideas' (Michael, 2017, p. 37), to borrow from work on key concepts in physiology, they must be unpacked for students to gain a clear idea of what they mean and how to use them. For example, as discussed in Chapter 5, the concept of space includes 11 different ideas—absolute location, relative location, distance, time-space convergence, accessibility, centrality, proximity, remoteness, spatial distribution, diffusion and the organisation of space—as well as four different ways of conceptualising space. Space, like the other core concepts, is consequently a simple word that covers many ideas, and all of these need to be understood before a student can adequately comprehend the meaning of space in geography. Much of this book discusses how each of geography's core concepts can be unpacked into more specific ideas that students can then apply, while Chapter 9 describes some of the strategies that teachers can use to develop student understanding of them.

If you have majored in geography at a university, you may have noticed that this view of geographical thinking is different to that in courses and books on geographic thought. These cover theoretical discourses such as spatial science, humanistic geographies, Marxist geographies, feminist geographies, postmodernism, poststructuralist geographies and relational geographies, to use Tim Cresswell's book on *Geographic thought: An introduction* (Cresswell, 2013) as an example. They strongly influence the different ways we think about and understand place, space, environment and interconnection, so that, for example, there are humanistic ways of understanding people's attachment to places and relational ways of explaining why places have their particular characteristics.

But they are not the core of geographical thinking, and they are not what differentiates geography from other school subjects.

These concepts have been rethought continuously as geographers have adopted new ways of viewing the world. The spread of Marxist thinking into the subject, for instance, changed the ways that space was conceptualised, while the introduction of feminist viewpoints changed thinking about place. Concepts are therefore contestable and evolving, and may have different meanings for different people in different contexts and at different times. The interpretations in this book are my own attempt to explain the core concepts in ways that are appropriate for teachers in schools and are therefore quite selective. My interpretations also try to explain, and sometimes critique, some contemporary thinking on them.

Geography's core concepts influence the practice of geography in a number of ways, from how geographers perceive the world to the design of public policies, and students should understand that they are not just big ideas but have functions that enable them to do geography as well as to think about it. The ways the core concepts are used in geography are introduced and briefly discussed below and will be further explained in the chapters on each one.

Ways of perceiving the world

Geography's key concepts provide distinctive ways of viewing and interrogating the world. For example, the processes and patterns of socioeconomic change as nations develop will be perceived differently by different disciplines. An economist is likely to focus on changes in the structure of the economy, a political scientist on changes in political institutions and a sociologist on changes in class structures, personal beliefs or gender relations. A geographer, on the other hand, is likely to study the causes and consequences of the spatial changes that both result from and contribute to national socioeconomic change, such as urbanisation, internal and international migration and the development of new economic regions and cities. In the study of health, a medical scientist might focus on the effects of individual characteristics such as age, sex and occupation on health outcomes, while a geographer might study the effects of the physical and social environment of the place in which people live (a place-based perspective), or of accessibility to health services (a spatial perspective), on their health. The ways of thinking of different disciplines consequently influence how they perceive and study the same phenomena.

Viewing geography as a way of thinking was central to the British Geographical Association's influential 2009 manifesto for school geography, *A different view*. The manifesto argued that 'One way of understanding geography is as a *language* that provides a way of thinking about the world: looking at it, investigating it, perhaps even understanding it in new ways' (Geographical Association, 2009, p. 10). These ways of thinking will be discussed in some depth in the chapters on the four key concepts. They are perhaps the most important of the types of powerful thinking discussed in this chapter, because they can completely change a student's perceptions of the world, what they think are significant issues, the questions they ask and even their behaviour.

Questioning

Questioning is educationally important because deep thinking is most likely to occur when students are trying to find an answer to a question, particularly one that does not have an obvious solution, because this requires them to analyse information, evaluate alternative possibilities and reach a reasoned and defensible conclusion. Learning the skill of questioning is also important because, as Phil Wood argues: 'If students are expected to develop independent learning skills, critical thinking capacities and, ultimately, the ability to carry out independent enquiry, then we need to model and develop the skill of questioning as a core concern' (Wood, 2006, p. 76). He therefore supports teaching strategies that give students a more active and independent role in the formulation of questions.

Generic questions, such as ones that start with 'what', 'where', 'why' and 'so what', are commonly used in geography, but they have been criticised for leading to descriptive responses that do not involve conceptual thinking, for being an inflexible approach to questioning (with the same questions being asked regardless of the context) and for lacking deep interest, especially if used repeatedly (Rowley, 2006). Margaret Roberts (2013) adds that while generic question frameworks can be helpful, they do not encourage curiosity and do not necessarily encourage geographical questions. Questions that are specific to the topic or issue being studied are more likely to be epistemically productive, particularly if they have these characteristics:

- They interest and engage students, and may have been generated by them.
- There is more than one possible answer, so students must evaluate the evidence for each.
- Students will be able to find the information needed to do this.
- They stimulate questioning, discussion and debate between students.
- The answer leads to thinking about what actions, if any, could or should be taken.

Chris Rowley also advocates questions that do not require the collection and analysis of data. He suggests that 'many areas of geography would benefit from an approach to enquiry which starts from values and ideas rather than from information collection; an approach where thinking and talking together replaces gathering and garnering information as the starting point' (Rowley, 2006, p. 21). This advice might lead to questions such as these:

- What do we like about our place?
- Is it useful to know where countries are?

- What makes a place interesting?
- Is sustainability about saving the planet or saving us?
- Is the inequality in incomes between countries fair?
- Do humans have an obligation to preserve other forms of life?

Rachel Lofthouse (2011) advocates the use of 'Big geographical questions' to provide a coherent framework for a range of studies. Her leading example is 'Why do people live where they do?', which could be the overall question for separate studies of the local area, a whole city, a country and migration at several scales. Another example is 'How and why does climate change from place to place and time to time?', which she suggests could be the overall question for studies of the world spatial distribution of climates, the effects of altitude on weather and climate and the consequences of global warming for different places. Both questions are based on the concepts of place and space and provide coherence for a variety of lessons and topics.

Questions are also generated by new ways of thinking. These are powerful because they lead to new questions, or at least ones that are new to a student, which in turn produce new knowledge and new insights. Distinctively geographical questions are derived from the subject's concepts. If students are studying a place, they might use the concept of space to ask the question 'are the characteristics of this place influenced by its relative location', while the concept of interconnection leads to the question 'are the characteristics of this place influenced by its relationships with other places', and the concept of environment prompts the question 'are the characteristics of this place influenced by its biophysical environment'. The role of concepts in questioning will be explored in later chapters.

Organising and analysing data

Before data can be analysed, they have to be organised in some way, and some of these ways are geographical. For example, data that are recorded and portrayed by place, such as by settlements, local government areas or environmental zones, applies the concept of place, while data that are mapped applies the concept of space. These are two very common ways of organising and portraying data in geography, and both are based on a key geographical concept.

Analysis is a term that can have a range of interpretations in schools and in everyday life. It comes from an Ancient Greek word meaning 'a breaking up', so analysis strictly means examining a complex topic by breaking it into its component parts, studying each of them separately and then putting them together again to identify their relationships. An example is explaining the volume of moisture in a soil by analysing the separate flows of water that determine it. This is described in Chapter 3 as an application of the concept of interconnection. Another type of analysis is to identify the essential features of something, such as the pattern in a table or map. The interpretation of spatial distributions is described in Chapter 5 as an application of the concept of space.

Comparison is also an analytical method when used to understand causal relationships. In geography it typically involves a comparison between places and therefore applies the concept of place, or a comparison of spatial distributions, which applies the concept of space. For many students, these will represent new ways of thinking.

Generalising

Generalisations produced from the study of individual cases or topics help students to consolidate and integrate their understanding. Geographical generalisations are based on the discipline's concepts, as in this example:

> Because of the advantages of proximity, economic activities tend to cluster in space, at all scales from the local to the global, unless tied to the location of natural resources or dispersed customers.[6]

The first part of this statement contains an explanation for the clustering of economic activities, while the last part identifies exceptions to the generalisation. Explanatory generalisations like these are powerful because they enable students to transfer their understanding to new contexts and to answer questions and solve problems that are new to them.

A generalisation is probably the nearest that geographers can get to a principle or law, but because of the uniqueness of places geographical generalisations may have many exceptions, particularly when human actions are involved. However, not all geographers agree that generalisation is possible. Those who follow a constructivist philosophy of knowledge argue that all knowledge is subjective and specific to a particular time and context and reject the possibility of generalisation. Geographers have also contended that the different characteristics of places alter the local outcomes of environmental and human processes and consequently limit generalisation. However, in this book, generalisations are encouraged, as long as it is recognised that they are not laws or principles, may apply to limited sets of situations and will have exceptions. Examining these exceptions, or anomalies, can help in understanding the relationships within a generalisation by revealing why, in this case, it does not apply, and this can be a productive learning exercise.

An example of the limitations on generalising comes from a study by the American physical geographer Jonathan Phillips of the effects of grazing and fire on the relative importance of grasses and trees/shrubs in three places in the southern USA. He discovered three different relationships between grazing, fire and vegetation, because of the different environmental characteristics of each place, and argues that 'Conditions specific to a location or region will influence what, how, or even if generalizations apply' (Phillips, 2001, p. 321).

André Roy and Stuart Lane (2003) have a similar view in their essay on fluvial geomorphology, in which they argue that generalisation about river behaviour is difficult because of the influence of context, by which they mean the uniqueness of the form of every meander bend or confluence in particular places, and how that form interacts with fluvial processes. Joseph Holden explains this point in more detail:

> For example, a meander bend may form at a given point only if: (i) there is a river flowing past the point; (ii) the right turbulent flows, sediment erosion and transport mechanisms are operating; (iii) the soil material is readily erodible to allow a meander bend to form; and (iv) local vegetation material or geology does not restrict the meander bend formation. In this case, the exact size, shape, and location of the meander bend will be a result of general meander-bend-producing processes and local factors.
>
> (Holden, 2017, p. 13)

On the other hand, a paper on river systems argues that:

> Recognizing that every location is potentially unique does not render generalizations meaningless. Regularities in time and space can still be observed as repeated patterns of landforms, and interpretations of these patterns can support efforts to meaningfully transfer understandings from one location to another. The challenge lies in identifying where a general pattern holds true and how and when local differences may be important. Theory (general understanding) informs local interpretations, but site-specific appraisals prompt insights into 'differences', allowing local 'stories' to emerge.
>
> (Brierley et al., 2013, p. 602)

Explaining

In education, the word 'explain' can have several meanings. Students are frequently asked to explain something, such as the meaning of a term or a statement, which requires them to demonstrate their understanding of it. A much more demanding task is to explain some phenomenon or event, which is called causal explanation to distinguish it from explanation as showing understanding. A causal explanation can be identified because it answers a why-question.

Patrick Meyfroidt (2016) explains that establishing causality relies on two dimensions: causal effects and causal mechanisms. A causal effect is when a change in a factor produces a change in an outcome. The factor, or combination of factors, is then thought to be the cause of the outcome. A complete causal explanation goes beyond this establishment of a relationship to identify how the factor (the cause) produces the change in the outcome. This is the causal mechanism. Meyfroidt illustrates the difference between a causal effect and a causal mechanism with the example of smoking and cancer. The causal effect of smoking on cancer was established statistically well before research discovered the causal mechanism that explained why smoking produced this outcome. Meyfroidt argues that both are needed for a strong explanation. A statement that Y always follows X is not an explanation until one can suggest how and why. Similarly, a statistical correlation between two variables does not mean that one has caused the other unless the causal link between them can be identified. Similarly, 'A plausible causal mechanism alone, without a corresponding causal effect, is nothing more than a mere conjecture or, at best, a hypothesis to test' (Meyfroidt, 2016, p. 504).

Explanation is not a simple task, because there may be several plausible explanations. Finding the best one could follow the steps recommended by Alec Fisher. These are:

1 Identify the possible explanations of the occurrence. This is to avoid the danger of assuming that there is only one possible cause, as while the evidence may suggest that this cause is plausible, others that have not been tested may be even better. It also avoids the trap of simply selecting the explanation or explanations favoured in the literature. The American geologist T. C. Chamberlin (1965) described this as the 'method of multiple working hypotheses', a way of guarding against attributing an explanation to a single cause. For each explanation, identify the cause-and-effect links, processes or reasons that are hypothesised to connect the explanation to the occurrence, because this is what explanation means.
2 Identify the evidence that would enable one to choose between these explanations.
3 Collect this evidence. As this is not always possible, students may need to be creative in thinking of alternative ways of testing relationships.
4 Decide which explanation is best supported by the evidence and best fits everything else we know that is relevant to this case. The second half of this step is about using one's existing knowledge and experience to help in making judgements. (Adapted from Fisher, 2011, pp. 153–159)

Explanation can also be complex because there may be more than one causal factor, with two or more needing to be present for the effect to be produced. Each factor alone is insufficient to produce the outcome that is being explained, as all are necessary for that outcome. For example, land clearing can produce erosion, but that outcome will also depend on soil type, slope and weather.

We can never be certain about explanations because, as discussed in Chapter 1, they are a product of human thought and are fallible and provisional. Causes usually cannot be directly observed but are attempts to make sense from what we can observe. This is the case with the natural sciences as well as the social sciences. Gravity, for instance, cannot be observed, but is a concept invented to explain observations. Geographers also have different approaches to identifying explanation, ranging from experimental methods and modelling in physical geography to an investigation of human actions or social structures in human geography. Furthermore, not everything in geography can be, or should be, 'explained', and qualitative studies in geography are usually more interested in interpretation, meaning and understanding than in explanation. For example, a person's sense of place, or feelings of belonging to the place in which they live (discussed in Chapter 4), is something we can try to understand but cannot easily explain.

An idea that is illustrated in some later chapters is that explanation can operate at successively deeper levels. For example, we could explain land degradation by identifying the environmental characteristics, such as climate and surface materials, that make the land susceptible to degradation. Or we could explain it by the actions of farmers in the ways they use the land. Or we could go further into the causal chain and ask why farmers engage in these actions, which in turn might be because they do not understand the consequences, or because inequalities in land ownership force them to overuse their land, or because they have been pushed onto marginal land unsuited to cultivation, or because communal sanctions on poor land use have broken down. An even deeper explanation would examine the structural social, economic and political causes of the explanations in the previous sentence. This way of thinking about explanation provides considerable scope for student debate.[7]

The explanations suggested in the previous paragraph range across the natural and social sciences, and for some topics could also include ideas from the humanities. This does not make them 'geographical explanations', because they are not derived from a key geographical concept. A geographical explanation might relate to the distinctive characteristics of a place (concept of place), relative location (concept of space) or the influence of the environment (concept of environment). Yet even these explanations are likely to be multi-disciplinary. The influence of relative location on businesses, for example, is through the costs and time involved in overcoming distance. Similarly, distance affects migration decisions through the psychological and economic costs of moving, and not just through physical distance, while the environment influences farming through the relative profitability of alternative crops, mediated by the culture and knowledge of the farmer. This wide-ranging way of explaining is one of the strengths of geography, because it produces a comprehensive analysis of causes and consequences, and not one limited to the perspective of a single discipline.

Applying

Applying or transferring knowledge to new situations and new problems is a particularly powerful form of thinking. Students can take what they have learned from one case study and apply their understanding to a new context. For example, they could apply their knowledge of the environmental, social and economic impacts of a particular flood to estimate the effects of flooding in a different location, or their knowledge of the effects of outmigration on a place to ask appropriate questions about the likely effects of outmigration on a different place. What is applied from one case study to another is either an understanding of the relationships between causes and outcomes or an awareness of the questions to ask. Note, however, that the usefulness of this method depends on whether the case to which knowledge is being transferred is sufficiently similar to the original case to make the transfer valid, and this similarity must be established before the original case study knowledge can be transferred. If what is transferred is an explanatory generalisation that successfully explains the new case study, this reinforces and verifies the power of the generalisation.

Forecasting

One of the quotations from Michael Young at the beginning of this chapter was that 'Knowledge is "powerful" if it … enables people to envisage alternatives'. This is thinking about futures, which can be very powerful if it leads students to identify the actions needed to achieve a preferred future and to think about how they could be involved in progress towards it. In school geography, probably the majority of topics studied have a potential futures dimension, ranging from local issues such as 'What might be the effects on this beach of a proposed coastal structure?' to regional issues such as 'What are the likely effects of climate change on this region?' and global issues such as 'What will be the effects of further land clearing on climate change?'

Forecasting is a term used here to describe the methods employed to think about futures, and the ones most relevant to school geography are described below.

Measuring past trends and projecting them into the future. This method might be appropriate for estimating the future population of a place or a group of places. The simplest method is to use Census data for past population numbers to establish a trend line that can be projected into the future. A more complex but more accurate method is to use current and estimated measures of births, deaths and migration to estimate future populations. This requires students to make assumptions about future levels of fertility, mortality and migration, which in turn depends on a good understanding of their causes. Another variant of the method would be to identify and explain past periods of growth or decline in population and use this understanding to create scenarios of future populations, based on assumptions about future causes of town growth or decline. Similarly, a study of future coastal erosion could start by estimating rates of erosion from historical satellite photos, and project these into the future. Relating past coastal erosion to storm events would refine this method by enabling students to incorporate assumptions about future storm intensity and frequency into their projections.

Applying a generalisation, which is a conceptual model, to construct scenarios of what might happen in the future. The following statement is an example of a generalisation that students might be able to make after a study of coastal processes in a particular locality:

> Coastal areas are dominated by wave and tidal processes that drive weathering and sediment movement, and stopping natural sediment movements in one place may cause erosion and sedimentation in other places.
>
> (Adapted from Holden, 2011, p. 119)

This is a statement of the relationships between processes and phenomena and not a description of facts. Generalisations such as this one are educationally useful because they help students to make sense of a lot of information and so increase their understanding. More importantly, 'they allow students to apply what they have learned to new settings and to transfer prior knowledge to new situations' (Shiveley and Misco, 2009, p. 76). The generalisation above, for example, could be used by students to forecast the effects of a proposed coastal structure, such as a marina or breakwater, using their understanding of the processes described in this statement. Another generalisation (to be explained in Chapter 3) that could help students to think about the future is:

> Because of the interconnections between the components of the biophysical environment, change in one component may produce change in others. The subsequent changes may be experienced in the same place as the initial change, and/or in different places, or at a different scale.

This guides students towards the questions they need to ask about the effects of an environmental change, such as the clearing of a forest or the damming of a river.

Imagining scenarios of the future, based on assumptions about technological, economic and social change. These assumptions could be based on the opinions of experts, the ideas of writers of science fiction or students' creative imaginations, as 'By using imagination, for example by means of divergent thinking ..., students can think beyond the fixed and familiar to arrive at novel, possible and preferable futures' (Pauw et al., 2018, p. 47). It is advisable to ensure that a diversity of opinions is sought from different types of people and not just from prominent experts.

Some geographical futures can be numerically predicted or forecast, but even in physical geography, the complexity and uncertainty of physical and biological systems makes this difficult, and even the best models are simplifications of the real world, based on assumptions that may turn out to be incorrect. Social phenomena are much harder to predict or forecast, because they are subject to unpredictable human actions and are generally better expressed as scenarios. Advocates of futures education in geography recommend that students develop several scenarios of the future of whatever issue or topic they are studying. These should be of three types:

- Probable (the most likely future based on current trends and knowledge).
- Possible (alternative futures that students can imagine).
- Preferred (the possible future that students would prefer).

The differences between them have been described by Alvin Toffler, a famous futurist, as being about 'visionary exploration of the possible, systematic investigation of the probable and moral evaluation of the preferable' (Bell, 2009, p. 73, as quoted in Pauw, 2015, p. 316). The inclusion of 'moral' is because probable and possible futures can be evaluated for their effects on environmental sustainability and human wellbeing, and a preferred future chosen for its contribution to them. Students can then use their understanding of the causes of different futures to think about what would be needed to achieve this preferred future. Importantly, making informed decisions on how to influence the future depends on a good understanding of causes and builds on the powerful thinking needed to generalise and explain. However, while the use of this knowledge is essential, 'futures education aims at taking the thinking process a fundamental step further, into imagination and envisioning' (Pauw, 2015, p. 317).

Evaluating

Evaluation can have several meanings. At one level it is a way of assessing the soundness of arguments and the validity of claims about knowledge and answers to questions. This takes several forms.

1 Evaluation of a question. Does the question contain assumptions that might affect the answers? Is it open to the possibility of different answers, or has the answer been pre-determined by the question?
2 Evaluation of a claim or answer. Is what is stated to be true based on facts or opinions? This will determine how the claim can be evaluated.
3 Are there alternative interpretations or perspectives, and how can they be evaluated? Dina Vasiljuk and Alexandra Budke (2012) have a complex but very useful discussion of what they call 'perspective-taking', where students are asked to examine 'issues or problems [where] there is no incontrovertible right or wrong opinion and solution, but rather different points of view which lead to different truths'. They go on to develop a model for teaching perspective-taking.
4 Evaluation of the information on which a claim or answer is based. Are the sources of the information credible? Were the methods used to collect the information likely to produce reliable data? Is there other information, not considered, that does not support the claim or answer?

5 Evaluation of the arguments. Are the reasons for the claim or answer logically argued? Do any causal explanations identify the relationships, processes or mechanisms that are thought to connect causes to outcomes? Where the arguments involve opinions and value judgements, can these be justified? Note that these are questions students should also ask of their own knowledge and not just of the claims of others.

At another level evaluation is thinking about the significance and meaning of what has been found out. Is it a matter for concern or simply another piece of knowledge? Is it something that should be addressed by policies and actions, or can it be noted and set aside? To answer these questions, the two evaluative concepts discussed in Chapter 8, sustainability and human wellbeing, can be applied to decide through discussion and debate whether what has been found out is a threat to the sustainability of vital environmental functions, the wellbeing of particular people or the future of a place. If the answer is that there is a threat, then the next step is to develop ideas on how to respond.

Responding

Responding requires students to use their knowledge to think of ways to respond to a problem they have identified. To do this effectively they must have a good understanding of the causes of the problem, think of strategies that will counter these causes and explain how any strategy will make a difference. Strategies to address environmental, economic and social issues often have a basis in one of geography's key concepts, although this may be implicit rather than explicit. For example, local and regional economic and social development strategies that address the unique problems and resources of a place are implicitly based on the concept of place. Similarly, the creation of vegetation corridors to link conservation areas illustrates the concept of interconnection, while strategies to develop isolated places by improving their transportation and communications connections with major centres, and consequently reducing the costs imposed by distance, illustrate the concept of space. This function of the key concepts is often missing from school geography and will be illustrated in later chapters.

Higher-order thinking

The discussion above has argued that geography's major concepts generate a variety of forms of thinking. Some are about ways of viewing and perceiving the world geographically and the questions to ask and investigate. Others are about ways of organising and analysing data; identifying and testing explanations; generalising one's understanding and applying that understanding to new problems and situations; evaluating the significance of observations and findings; and selecting public policies and strategies to address environmental and social issues.

Teachers will recognise that many of the types of thinking described above are examples of higher-order and critical thinking. While some educators contend that these thinking skills can be taught on their own, others argue that they are closely linked to the subject matter and ways of thinking of a discipline and cannot be separated from disciplinary knowledge. As Sharon Bailin and her co-authors argue:

> critical thinking always takes place in the context of (and against the backdrop of) already existing concepts, beliefs, values, and ways of acting. This context plays a very significant role in determining what will count as sensible or reasonable application of standards and principles of good thinking. Thus, the depth of knowledge, understanding and experience persons have in a particular area of study or practice is a significant determinant of the degree to which they are capable of thinking critically in that area.
>
> (Bailin et al., 1999, p. 290)

This may be one reason why a review of studies of critical thinking interventions in US colleges concluded that 'interventions targeting general critical thinking skills and dispositions were only moderately effective, but discipline-specific critical thinking interventions were more promising' (Greene and Yu, 2016, p. 46). Similarly, Catherine Little (2018, p. 379) reports that research has 'highlighted that thinking skills do not necessarily transfer automatically for learners from an isolated thinking skills program to other contexts' and recommends approaches that combine content-embedded critical thinking with explicit instruction in critical thinking methods. This book embeds critical thinking within the discipline of geography.

Note that the types of thinking discussed in this chapter, and in the rest of the book, go beyond higher-order thinking. One addition is discipline-specific ways of perceiving or thinking about the world, which comes from the concept of powerful knowledge. These are based on a subject's core concepts, and explaining them is a major part of the book. The second is the use of the concepts of sustainability and human wellbeing to evaluate environmental and social trends for their ethical consequences. This produces a deeper form of evaluation.

A benefit of incorporating the thinking abilities described in this chapter into school geography is that they can make the subject more interesting, because they reduce the emphasis on remembering quantities of factual information and provide more scope for investigating real-life problems, debating answers and making judgements. This makes students active participants in their learning rather than passive recipients of information and geography a more interesting and intellectually challenging subject.

Can these types of thinking be learned by all students?

An important question is whether some of these types of thinking are appropriate for students perceived as lower performing. Research in the USA, for example, finds that many teachers refrain from assigning higher-order thinking tasks to lower-performing students because it is assumed they will find them hard and frustrating. This then limits the intellectual development of these students and denies them the opportunity to improve. On the other hand, a review of four programs designed to develop the higher-order thinking of all students found that:

> Students with both high and low academic achievements gained significantly from the educational interventions. Contrary to many practitioners' beliefs ... our empirical evidence shows that instruction of higher order thinking skills is appropriate for students with high and low academic achievements alike.
>
> (Zohar and Dori, 2003, p. 174)

A second issue is at what ages are students able to engage in powerful thinking? A review of a range of research studies on critical-analytic thinking (CAT), another term for critical thinking, came to these conclusions:

- 'children as young as 3 years of age are capable of abstract thought and have some of the thinking abilities that are needed for CAT in certain situations'
- 'young children, and even perhaps infants, can take the perspectives of others into account'
- 'children between ages 7 and 10 continue to show increased insight into argument quality'
- 'young children have a surprising sensitivity to the trustworthiness of different sources of information'
- 'although children below the age 7 may have difficulty with some aspects of CAT, they could nevertheless gain proficiency in project-based learning approaches' (Byrnes and Dunbar, 2014, pp. 285 and 290).

These findings suggest that thinking skills can be incorporated into teaching and learning from early in primary school, with appropriate methods and scaffolding.

Conclusion

Geography's core concepts, as identified in this chapter, help students to (1) think geographically, through the ways of perceiving the world that they produce; (2) deeply, through the ways of thinking they support; and (3) ethically, through their ways of evaluating. Teaching that focuses on developing students' understanding of geography's core concepts has other benefits. These concepts make geography a distinctive discipline and give it coherence despite its enormous range in subject matter. They are essential for intellectual development because this is based on conceptual thinking. They can also help to overcome information overload by following Harlen's advice that the goal of science education 'is not knowledge of a body of facts and theories but a progression towards key ideas which enable understanding of events and phenomena of relevance to students' lives' (Harlen, 2010, p. 2).

Summary

- The concept of powerful knowledge provides a way of identifying some important types of thinking.
- Geographical thinking is based on its concepts.
- Geography's four key concepts are place, space, environment and interconnection, each of them at the top of a hierarchy of increasingly abstract and complex ideas.
- Scale and time are analytical concepts, and sustainability and human wellbeing are evaluative concepts.
- The key concepts must be unpacked to be useable by students.
- All concepts are contestable, and ideas about them have changed over time.
- The key concepts have a number of functions, such as providing distinctive ways of perceiving the world, identifying questions to ask, organising information, suggesting methods of analysis, forming generalisations, identifying possible explanations, forecasting futures, evaluating outcomes and informing public policies.
- They also teach higher-order and critical thinking, but through the discipline.
- The concepts make geography a distinctive and coherent discipline and are essential to the intellectual development of students.

How could you use this chapter in teaching?

The ways of thinking discussed in this chapter can be developed in any of the topics studied in school geography, and Chapters 10 and 11 provide examples of their application to two teaching topics. They should not be taught on their own, and instead students should gain an understanding of the different types of thinking and how to use them through their study of a range of substantive topics. However, this will depend on the extent to which teaching is designed to develop thinking and whether it follows some of the suggestions in Chapter 9.

Notes

1 For example, Dolan (2020), and Enser (2021).
2 For a fuller discussion, see Maude (2016 and 2017).
3 Some of this material was first published in Maude (2021). This publication also lists the academic and educational sources used to identify candidates for geography's core concepts.
4 Gregory and Lewin (2018) have constructed detailed hierarchies for six multi-disciplinary meta-concepts and their subsidiary concepts in physical geography. While written for university teachers and researchers, and going well beyond the scope of this chapter, teachers in schools may find their article useful.
5 See Dessen Jankell et al. (2021) for another way of selecting and organising geography's concepts. It identifies a broadly similar list of concepts but arranges them differently.
6 Statements like these, unless otherwise acknowledged, are by the author.
7 See also the section on land degradation in Chapter 6.

Useful readings

Hanson, S. (2004). Who are 'we'? An important question for geography's future. *Annals of the Association of American Geographers, 94*, 715–722. https://doi.org/10.1111/j.1467-8306.2004.00425.x

An insightful discussion of a 'set of uniquely geographic propositions that is at the core of our work as geographers', by a leading American geographer.

Heffron, S. G., & Downs, R. M. (2012). *Geography for life: National geography standards*. National Council for Geographic Education.

This major US document on the content of geography in schools identifies the spatial and ecological perspectives as the key ones in geography, and these are similar to the concepts of space, environment and interconnection in this book.

Jackson, P. (2006). Thinking geographically. *Geography, 91*, 199–204. https://doi.org/10.1080/00167487.2006.12094167

A leading British geographer's ideas on key geographical concepts.

Lambert, D. (2017). Thinking geographically. In M. Jones (Ed.), *The handbook of secondary geography* (pp. 20–29). Geographical Association.

A short article by a leading geography educator that presents a sequence of views on geography as a discipline, and then outlines the British Geographical Association's framework for 'thinking geographically', based on the concepts of place, space and environment. This is probably the best short English-language account of geographical thinking, and many of its ideas are further developed in this book.

Lambert, D., & Morgan, J. (2010). *Teaching geography 11-18: A conceptual approach*. Open University Press.

Good discussion of ideas on most of the concepts included in this book.

Murphy, A. B. (2018). *Geography: Why it matters*. Polity Press.

Readable and informative chapters on geography and space, place and environment, by a leading American geographer.

Roberts, M. (2013). Developing conceptual understanding through geographical enquiry. In M. Roberts (Ed.), *Geography through enquiry: Approaches to teaching and learning in the secondary school* (pp. 81–94). Geographical Association.

A useful discussion of concepts and how inquiry methods can be designed to support conceptual development.

References

Alderson, P. (2020). Powerful knowledge, myth or reality? Four necessary conditions if knowledge is to be associated with power and social justice. *London Review of Education*, *18*(1), 96–106. https://doi.org/10.18546/LRE.18.1.07

Bailin, S., Case, R., Coombs, J. R., & Daniels, L. B. (1999). Conceptualizing critical thinking. *Journal of Curriculum Studies*, *31*(3), 285–302. https://doi.org/10.1080/002202799183124

Bell, W. (2009). *Foundations of futures studies, volume 1: History, purposes and knowledge*. Transaction.

Brierley, G., Fryirs, K., Cullum, C., Tadaki, M., Huang, H. Q., & Blue, B. (2013). Reading the landscape: Integrating the theory and practice of geomorphology to develop place-based understandings of river systems. *Progress in Physical Geography*, *37*, 601–621. https://doi.org/10.1177/0309133313490007

Brooks, C. (2018). Understanding conceptual development in school geography. In M. Jones & D. Lambert (Eds.), *Debates in geography education* (2nd ed., pp. 103–114). Routledge.

Byrnes, J. P., & Dunbar, K. N. (2014). The nature and development of critical-analytic thinking. *Educational Psychology Review*, *26*(4), 477–493. https://doi.org/10.1007/s10648-014-9284-0

Chamberlin, T. C. (1965). The method of multiple working hypotheses. *Science*, *148*(3671), 754–759. (Originally published in 1890). https://doi.org/10.1126/science.148.3671.754

Cresswell, T. (2013). *Geographic thought: An introduction*. Wiley-Blackwell.

Dessen Jankell, L., Sandahl, J., & Örbring, D. (2021). Organising concepts in geography education: A model. *Geography*, *106*(2), 66–75. https://doi.org/10.1080/00167487.2021.1919406

Dolan, A. M. (2020). *Powerful primary geography: A toolkit for 21-st century learning*. Routledge.

Enser, M. (2021). *Powerful geography: A curriculum with purpose in practice*. Crown House.

Fisher, A. (2011). *Critical thinking: An introduction*. Cambridge University Press.

Fögele, J. (2017). Acquiring powerful thinking through geographical key concepts. In C. Brooks, G. Butt, & M. Fargher (Eds.), *The power of geographical thinking* (pp. 59–73). Springer.

Geographical Association. (2009). *A different view: A manifesto from the Geographical Association*. Geographical Association.

Geographical Association. (n.d.). *Concepts in geography*. Geographical Association.

Greene, J. A., & Yu, S. B. (2016). Educating critical thinkers: The role of epistemic cognition. *Policy Insights from the Behavioral and Brain Sciences*, *3*(1), 45–53. https://doi.org/10.1177/2372732215622223

Gregory, K. J., & Lewin, J. (2018). A hierarchical framework for concepts in physical geography. *Progress in Physical Geography*, *42*, 721–738. https://doi.org/10.1177/0309133318794502

Gregory, K. J., & Lewin, J. (2021). Big ideas in the geography curriculum: Nature, awareness and need. *Journal of Geography in Higher Education*, *47*(1), 9–28. https://doi.org/10.1080/03098265.2021.1980867

Harlen, W. (Ed.) (2010). *Principles and big ideas of science education*. Association for Science Education.

Holden, J. (2011). *Physical geography: The basics*. Routledge.
Holden, J. (Ed.) (2017). *An introduction to physical geography and the environment* (4th ed.). Pearson.
Lambert, D., Solem, M., & Tani, S. (2015) Achieving human potential through geography education: A capabilities approach to curriculum making in schools. *Annals of the Association of American Geographers*, *105*(4), 723–735. https://doi.org/10.1080/00045608.2015.1022128
Little, C. A. (2018). Teaching strategies to support the education of gifted learners. In S. I. Pfeiffer (Ed.), *APA handbook of giftedness and talent* (pp. 371–385). American Psychological Association.
Lofthouse, R. (2011). Is this big enough? Using big geographical questions to develop subject pedagogy. *Teaching Geography*, *36*(1), 20–21. https://portal.geography.org.uk/journal/view/J004513
Maude, A. (2016). What might powerful geographical knowledge look like? *Geography*, *101*(2), 70–76. https://doi.org/10.1080/00167487.2016.12093987
Maude, A. (2017). Applying the concept of powerful knowledge to school geography. In C. Brooks, G. Butt, & M. Fargher (Eds.), *The power of geographical thinking* (pp. 27–40). Springer.
Maude, A. (2021). Recontextualisation: Selecting and expressing geography's 'big ideas'. In M. Fargher, D. Mitchell, & E. Till, (Eds.), *Recontextualising geography in education* (pp. 27–40). Springer. https://link.springer.com/chapter/10.1007/978-3-030-73722-1_3
Meyfroidt, P. (2016). Approaches and terminology for causal analysis in land systems science. *Journal of Land Use Science*, *11*(5), 501–522. https://doi.org/10.1080/1747423X.2015.1117530
Michael, J. (2017). What does it mean to 'unpack' a core concept? In J. Michael, W. Cliff, J. McFarland, H. Modell, & A. Wright (Eds.), *The Core concepts of physiology* (pp. 37–44). Springer.
Mitchell, D., & Stones, A. (2022). Disciplinary knowledge for what ends? The values dimension in curriculum research in a time of environmental crisis. *London Review of Education*, *20*(1), 23. https://doi.org/10.14324/LRE.20.1.23
Pauw, I. (2015). Educating for the future: The position of school geography. *International Research in Geographical and Environmental Education*, *24*(4), 307–324. https://doi.org/10.1080/10382046.2015.1086103
Pauw, I., Béneker, T., van der Schee, J., & van der Vaart, R. (2018). Students' abilities to envision scenarios of urban futures. *Journal of Future Studies*, *23*(2), 45–65. https://doi.org/10.6531/JFS.201812_23(2).0004
Phillips, J. D. (2001). Human impacts on the environment: Unpredictability and the primacy of place. *Physical Geography*, *22*(4), 321–332. https://doi.org/10.1080/02723646.2001.10642746
Rawling, E. (2007). *Planning your key stage 3 geography curriculum*. Geographical Association.
Rowley, C. (2006). Are there different types of geographical inquiry? In H. Cooper, C. Rowley, & S. Asquith (Eds.), *Geography 3–11: A guide for teachers* (pp. 17–32). David Fulton.
Roy, A., & Lane, S. (2003). Putting the morphology back into fluvial geomorphology: The case of river meanders and tributary junctions. In S. Trudgill & A. Roy (Eds.), *Contemporary meanings in physical geography: From what to why?* (pp. 103–125). Hodder Education.

Shiveley, J., & Misco, T. (2009). Reclaiming generalizations in social studies education. *Social Studies Research and Practice*, *4*(2), 73–78. https://doi.org/10.1108/SSRP-02-2009-B0006

Vasiljuk, D., & Budke, A. (2021). Multiperspectivity as a process of understanding and reflection: Introduction to a model for perspective-taking in geography education. *European Journal of Investigation in Health, Psychology and Education*, *11*(2), 529–545. https://doi.org/10.3390/ejihpe11020038

Wood, P. (2006). Developing enquiry through questioning. *Teaching Geography*, *31*, 76–78. https://portal.geography.org.uk/journal/view/J004878

Young, M. (2008). From constructivism to realism in the sociology of the curriculum. *Review of Research in Education*, *32*, 1–32. https://doi.org/10.3102/0091732X07308969

Young, M. (2013). Powerful knowledge: An analytically useful concept or just a 'sexy sounding term'? A response to John Beck's 'Powerful knowledge, esoteric knowledge, curriculum knowledge'. *Cambridge Journal of Education*, *43*, 195–198. https://doi.org/10.1080/0305764X.2013.776356

Young, M. (2014). Knowledge, curriculum and the future school. In M. Young & D. Lambert with C. Roberts & M. Roberts (Eds.), *Knowledge and the future school: Curriculum and social justice* (pp. 9–40). Bloomsbury.

Young, M. (2015, September 1). Unleashing the power of knowledge for all. *Spiked*. https://www.spiked-online.com/2015/09/01/unleashing-the-power-of-knowledge-for-all/

Zohar, A., & Dori, Y. J. (2003). Higher order thinking skills and low-achieving students: Are they mutually exclusive? *The Journal of the Learning Sciences*, *12*(2), 145–181. https://doi.org/10.1207/S15327809JLS1202_1

Chapter 3

Thinking geographically

Interconnection

The concept of interconnection is about recognising that phenomena do not exist in isolation but are always linked to other phenomena in complex ways. It is discussed before place, space and environment because it has terms and ideas that will be applied in the chapters on those concepts and that are frequently used in conjunction with them. The concept of systems, for example, is explained in this chapter but is also an essential idea for understanding the environment. Interconnection is a high-level concept that incorporates a number of subsidiary ideas. The ones discussed in this chapter are:

- Interdependence
- Flows
- Processes
- Systems
- Causal relationships
- Relational thinking
- Holistic thinking

Interdependence

Interdependence is about the mutual dependence of phenomena on each other. It is used in physical geography to describe the mutual dependence of organisms in biological systems. For example, in a tropical forest, plants depend on nutrients in the soil to grow, as explained in more depth later in this chapter. Herbivores eat the plants and carnivores eat the herbivores. Dead animals and plant materials are decomposed by microorganisms on the forest floor, and this returns nutrients to the soil. The plants depend on the microorganisms to recycle nutrients, and the microorganisms depend on plants to supply organic matter; they are interdependent. A human geography example of interdependence is that between urban and rural areas. Historically the growth of towns and cities depended on the ability of rural areas to produce surpluses to sustain urban populations, as discussed later in this chapter. Today, rural

DOI: 10.4324/9781003376668-4

areas depend on urban areas for markets for some of their products, supplies of some of their inputs and a wide range of specialised services. In urbanising countries some rural areas also partly depend on remittances from urban migrants. Urban areas, on the other hand, depend on rural areas for food, water, raw materials (such as timber, sand and metals), energy (from oil, coal and gas, and from wind, solar and hydroelectric power) and places for recreation, holiday homes and retirement.

In school geography, the concept of interdependence is often used in relation to international trade, to describe the dependence of producers and consumers on each other. Students learn about their connections with other places and people in the world through the products that they buy, from food items to electronic equipment. It is often linked to Fair Trade and aspirations such as global citizenship, but the extent to which these relationships through trade can be described as mutually interdependent is debatable. In many such relationships, the producers are dependent on the buyer, but the buyer may be able to obtain the same product from several sources and is consequently not dependent on the one producer. This situation is illustrated by the following story.

> [A]fter a collision of Chinese and Japanese ships near the disputed Senkaku/Diaoyu Islands in the East China Sea in 2010, China punished Japan by restricting exports of rare-earth metals, which are essential in modern electronics. The result was that Japan lent money to an Australian mining company with a refinery in Malaysia, which today meets nearly one-third of Japanese demand. In addition, the Mountain Pass mine in California, which had shut in the early 2000s, was reopened. China's share of global rare-earth production has fallen from more than 95 percent in 2010 to 70 percent over a half dozen years. Short-term manipulation of interdependence encouraged the development of alternatives to reduce long-term vulnerability.
>
> (Nye, 2020, pp. 8–9)

As an aspiration, however, the concept of interdependence is very powerful, because it encourages people to think of the ethics of their relationships with other places and other people, such as through trade, migration, tourism, pollution, waste disposal and carbon emissions.

Flows

Flows between places are an obvious type of interconnection. They include ocean currents, tides, atmospheric circulation, water in a river basin, sediments in a tidal estuary, trade in commodities and the movement of capital. Their significance is in their effects on the places they flow from and the places they flow to, because they change both but in different ways.

Figure 3.1 shows the flows of water that determine the moisture content of a soil. Soil moisture is often called 'green water' to distinguish it from the 'blue water' in rivers, lakes and dams, and it supports most of the world's agricultural production. Moisture flows into the soil through rainfall and snowmelt, and is lost to the soil through transpiration from plants, evaporation from the soil, runoff across the land surface and flows to groundwater and streams. To understand the changing volume of water in a soil, each of these flows needs to be estimated separately and then combined to produce a measure of soil moisture change, known as the soil moisture budget. This budget is an important concept, because it explains the moisture available for the growth of vegetation and crops and affects both farmers and gardeners. David Legates et al. (2011, p. 65) argue that it can be an integrative theme in physical geography, because soil moisture is 'an important variable in regional and microclimatic analyses, landscape denudation and change through weathering, runoff generation and partitioning, mass wasting, and sediment transport'.

A type of water flow that may be unfamiliar to teachers and students is the flow of water vapour through the atmosphere. Water that evaporates into the atmosphere is often regarded as a loss, because it is no longer available for human use. However, this water will eventually become precipitation somewhere, and is an essential part of the water cycle. It is estimated that about 40% of precipitation over land comes from moisture derived from evaporation from vegetation and soil. Agricultural expansion and land cover change are reducing these flows, with actual and potential negative consequences for environments and humans (Keys et al., 2019). This issue will be further discussed in Chapter 10.

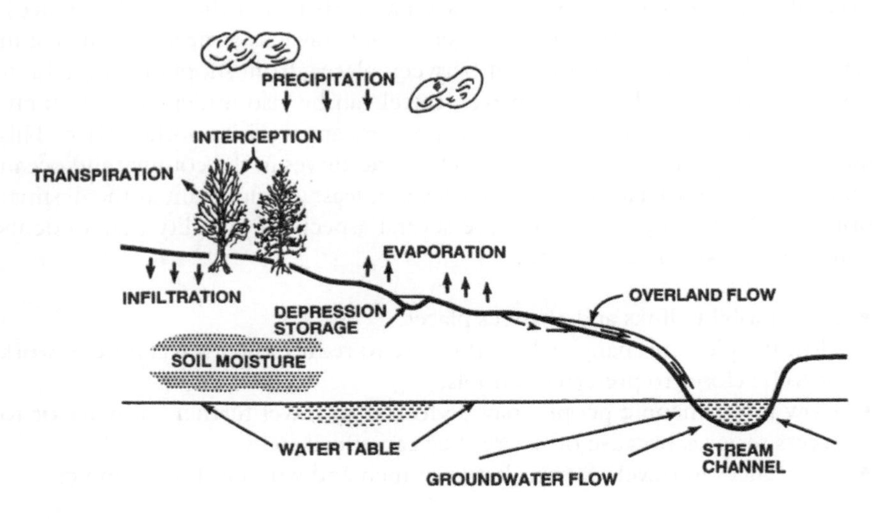

Figure 3.1 Flows that determine soil moisture.

Source: Adapted from Wetzel (2001), p. 44. Copyright Elsevier.

Human mobility is another example of flows. The term mobility is used to describe the great variety of types of human movement from one place to another, and encompasses far more than migration. Pooley identifies a continuum of forms of mobility from short-distance everyday moves to less frequent long-distance migration:

- Everyday movement around the home and garden.
- Daily short-distance movement for school, work, shopping, family and social activities, leisure and pleasure.
- Regular longer trips for business, family, social, leisure and other activities.
- Cyclical mobility (weekly, monthly, yearly) between two homes (long-distance weekly commuting; students moving between home and university; children moving between two parents).
- Holidays away from home.
- Local residential moves within the same community.
- Longer-distance migration within the same country.
- International migration. (Pooley, 2009, p. 144)

Pooley adds to these the concept of virtual mobility, which is about how the 'development of the Internet, mobile phones, and related mobile communications allow people to interact and be "mobile" without physical movement or face-to-face contact' (Pooley, 2009, p. 144). This idea is further discussed in Chapter 5. Note that people's mobility is affected by who they are. It may be restricted by physical disability, age, reliance on public transport, commitments (such as caring others), fear (of crime, assault or exposure to an infectious disease) and control by others (such as husbands or the morality police).

Studies of mobility have been largely about travel as a means to engage in activities which require movement between places, from shopping and education to work and holidays. However, travel can be also undertaken as an enjoyable experience in itself and not just as a means to visit another place. This idea might apply to walking as a recreation, car drives in the country and ocean cruises, as in these examples, the journey is at least as important as the destination. In school geography, there are several aspects of mobility that students could investigate. These include:

- How mobility links and changes places.
- How people may change where they live to reduce their travel time to work or to be closer to preferred schools.
- How lower-income people may be forced to travel further for work or to access services because of where they can afford to live.
- Differences in travel patterns between men and women. For example:

  Feminist geographers have documented how men tend to travel further to work but also tend to have simpler journeys which are less reliant on public transport. They do not have to pick up the children or do family shopping as often as women. (Creswell, 2011, p. 578)

Processes

'Process' is a term that is widely used in school geography but rarely explained, so this section tries to clarify its meaning and use. A process is a sequence of actions or events that determine the characteristics of something, such as the shape of a beach, the channel of a river, the migration of people or the growth of cities. Processes are sequences that are repeated in different places and over time and are generic rather than singular occurrences. These actions or events are sequentially connected as chains of cause-and-effect relationships, which is why 'process' belongs to the concept of interconnection. The process of coastal erosion, for example, occurs when atmospheric conditions produce strong winds that in turn create high energy waves which then erode and transport coastal materials and deposit them somewhere else. In chain migration, the initial movement of people to a new place may, through the flow of information and often money back to the place of origin, encourage and finance further migration over a number of years, so there is again a sequence of cause-and-effect relationships. The names given to processes are either descriptions of the process, such as urbanisation or weathering, or of the agent that drives the process, such as aeolian processes driven by the action of wind. The examples in this section also illustrate that processes operate over durations of time from very short to very long.

Most processes produce change, such as the erosion and retreat of coastal features, but some maintain the phenomenon that they produce. In chain migration, for example, once started the interconnections in the process may maintain the pattern for several decades. Similarly, prevailing wave patterns may maintain the shape of a beach over many years.

Processes are descriptions of change, not explanations. In physical geography, the explanations implicit in processes are the physical, chemical and biological mechanisms that produce change, except where humans are involved. Weathering, for example, is produced when rock is broken into smaller fragments by mechanisms such as freezing and thawing, heating, chemical solution and penetration by the roots of plants. In human geography, on the other hand, the explanations lie in the decisions and actions of people, businesses, institutions and governments, and the contexts that influence these decisions, and these can also be described as mechanisms. As Andrew Vayda et al. (1991, p. 320) caution: 'Social scientists sometimes use process to explain reality as if processes had lives of their own, as if they existed independent of human agency and were regulated by some larger dynamic in history'.

The importance of being able to identify the causal mechanisms, as explained in Chapter 2, is that without understanding them it is not possible to explain change, as they determine whether change happens and the form it takes. For example, the rate of weathering depends on the composition of the materials being weathered and the conditions that produce the weathering mechanisms, such as rainfall. The decision to migrate is not only a simple response to the existence of better economic opportunities in another place but also depends on whether people have the skills to make a living in that place, the resources to pay the cost of moving and a belief that their lives can be improved through migration.

The terms 'process' and 'mechanism' are often used interchangeably, which may be a source of confusion, but the distinction between them can be illustrated by unpacking the process of urbanisation. The term 'urbanisation' refers to an increasing proportion of the population of a region or country living in urban areas. In school textbooks, urbanisation is usually explained as a result of migration from rural to urban areas, in turn caused by the growth of jobs in manufacturing and services which tend to concentrate in towns and cities, and sometimes by rural poverty that motivates people to migrate. The emphasis is therefore on the effects of structural economic change on the location of employment and economic opportunities, and it is consequently difficult to explain urbanisation in countries which have not experienced the same economic transitions as those classed as 'developed'. An alternative interpretation comes from historical studies and the more recent experience of Africa. Sean Fox (2012) argues that the growth of pre-industrial cities was limited by their high mortality, and the lack of a surplus of food and fuel to support large urban populations, and not by the absence of actual or potential rural-urban migrants. The development of technologies and institutions that controlled disease, raised agricultural productivity, reduced transport costs and developed new sources of energy removed these limitations and allowed urban areas to grow through an excess of births over deaths as well as by migration from rural areas. Economic development could contribute to urbanisation if it increased non-agricultural employment, but urbanisation would occur in these places even if economic growth was poor because the limitations on urban growth had been removed. Fox therefore argues that there are three mechanisms that influence the process of urbanisation—disease control, food and energy surpluses and economic development—and the relative strength of each influences the characteristics of the towns and cities that they produce.[1]

Systems

A system is a set of interconnected objects that function together and can be conceptualised as an integrated whole. As G. P. Chapman describes, 'The word system is used to describe those phenomena where the parts are related to the whole in a manner which suggests that there are some organizing principles involved in the assembly of the whole' (Chapman, 2009, p. 146). Systems can be analysed through their three components:

- The objects forming the system.
- The relationships or interconnections between those objects.
- The interconnections between the system and its surrounding environment.

A systems approach is widely used in physical geography to understand whole environmental units such as ecosystems and drainage basins. Of these, ecosystems are perhaps the most common type studied in school geography and form a hierarchy that builds from small vegetation communities to major world biomes. Why they are called systems can be illustrated from the tropical rainforest (Figure 3.2). The components of the rainforest system are soil, vegetation, animals and forest floor litter (fallen leaves, branches, animal droppings and dead animals). The main interconnection between them is through the flow of plant nutrients. Plants take up nutrients from the soil and use them for growth. When leaves, branches and whole trees fall, and when the animals who feed on the leaves, or on other animals, deposit their droppings or die, nutrients are added to the litter. As this litter decomposes, it returns nutrients

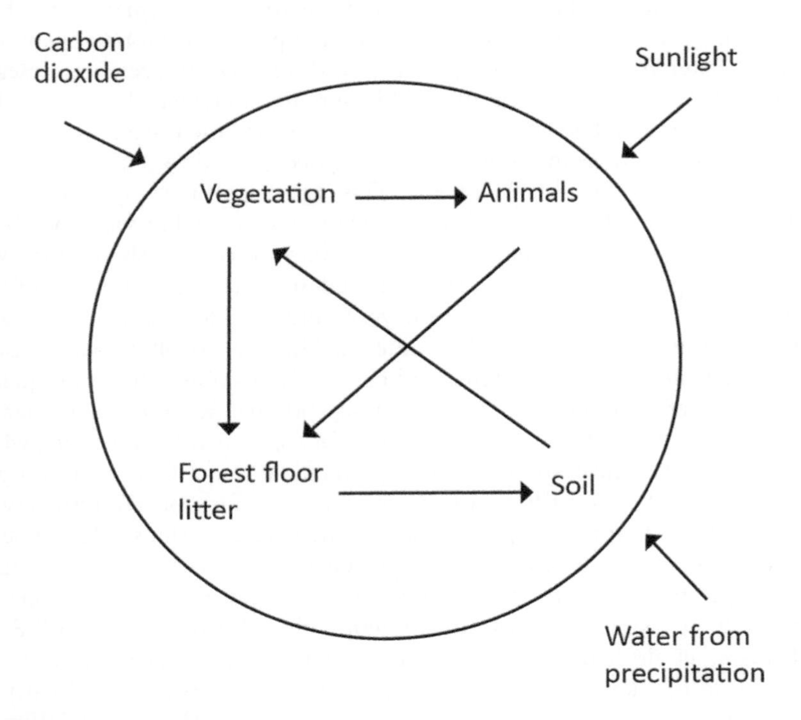

Figure 3.2 The rainforest system.

The circle is the boundary of the system. The arrows within the circle represent flows of plant nutrients.

Source: Author.

to the soil, where they can be taken up by plants in a continuing cycle. The system consequently works to maintain the whole rainforest ecosystem. However, if the vegetation is cleared, the cycle is broken, and the soil rapidly loses nutrients as they are leached out by the high rainfall of the humid tropical environment.

The third component of a system is its interconnection with the surrounding environment. Systems are closed if the objects in them do not interact with anything outside the system and open if they do; in geography, both physical and human, all systems are open because of the interconnectedness of the world. The tropical forest illustrates the concept of an open system because the components in the forest ecosystem have connections that go outside the boundaries of the system. For example, water from rainfall and carbon dioxide from the atmosphere are combined with sunlight through the process of photosynthesis to produce the sugars needed for plant growth and repair, and all three of these inputs are from outside the forest ecosystem.

In environmental systems, the objects are connected through flows of energy and materials. However, it is tempting to add information to these, because of the research which suggests that even plants have forms of communication. Peter Wohlleben (2015), in *The hidden life of trees*, refers to research that finds that in Africa acacias attacked by giraffes emit a gas that warns other trees of the danger. The latter then pump toxins into their leaves to deter the giraffes. Similarly, trees in other regions have been found to send warning signals to other trees through their roots. This is perhaps an example of how our beliefs about the differences between humans and other life forms are being slowly eroded as we learn more about the behaviour of animals and plants.

The concept of a system has also been used in human geography, but seems to have gone out of favour in recent years and very few human geography textbooks now list 'systems' in their indexes. This is probably for two reasons. One is that most human 'systems' are extremely open and are consequently difficult to define, draw boundaries around and analyse. Another is that the concept may be seen to be too mechanistic for topics involving people, whose actions are not sufficiently uniform or predictable. Noel Castree also argues that mechanistic systems frameworks do not allow for non-Western ways of thinking about nature and people-environment relationships. He advocates a range of visions that 'combine the evidence and forecasting produced by physical geographers with the concerns about values, goals, institutions and social systems characteristic of human geography' (Castree, 2015, p. 73). In such cases, an alternative conceptual focus is the idea of causal relationships discussed in the next section. Physical geography also has problems with systems, with research finding that many are extremely complex and their dynamics non-linear and unpredictable.

While formal systems analysis may be inappropriate for many geographical topics, the idea of a system reinforces the value of thinking about the relationships between things. It encourages relational thinking and holistic thinking, both of which are discussed later in this chapter. As Debra Straussfogel and Caroline von Schilling write:

> systems thinkers are expansionists, who would seek to understand a system by examining its processes and relationships as opposed to the reductionists of classical science, who would seek to understand a system by first examining its elementary parts and then by adding them together.
>
> (Straussfogel and von Schilling, 2009, p. 151)

Systems thinking is also useful when trying to understand or predict change. Because of the interconnections between the components of the system, change in one produces change in others. The subsequent changes may be experienced in the same place as the initial change, and/or in different places, or at a different scale, or at a different time. To again use the tropical rainforest as an example, if the vegetation is cleared and not replaced, the outcomes could be:

- Loss of soil nutrients (a change in the same place).
- Soil erosion and deposition of sediments downstream (a change in a different place).
- Future global warming because of the loss of a store of carbon dioxide (a change at a different scale and different time).

This example illustrates how a systems approach can involve the four concepts of interconnection, place, scale and time and produce a very geographical way of thinking.

An important aspect of systems thinking is the concept of feedback. This refers to the influence that an effect may have on its cause. In positive feedback, the influence acts to increase the effect, while negative feedback reduces the effect. An example of negative feedback comes from the actions of storms on beaches. Sediments eroded from the beach in a storm may be deposited as an offshore bar, which will reduce the energy of waves reaching the beach and reduce further beach erosion until the bar is removed by continued wave action. An example of positive feedback comes from chain migration. If migrants to a city are successful, the information and money they send back to their places of origin will have a positive effect on further migration. Both these examples could be described as systems because it is possible to identify the components that are interconnected, draw boundaries around them and describe the interconnections between them. However, they differ in that the first is a set of objects linked by physical flows of energy and materials, while the second is a set of objects (places or communities) linked by flows of people, information and money.[2]

Research finds that students can have difficulty with some aspects of systems, in particular with recognising that some relationships within a system are two-way, not linear, and that there are interactions between humans and the biophysical environment. Cox et al. (2019) recommend that teachers have students construct causal diagrams (sometimes also called concept maps) to visually represent the relationships involved in a particular situation or problem, which could be deforestation, flooding, refugee migration or racial segregation. The article by Dessen Jankell and Johansson listed at the end of this chapter also has ideas on how to teach systems thinking.

Causal relationships

Causal relationships are interconnections in which something is caused, at least partially, by something else, so that the relationship between them is one of cause and effect. In geography, these relationships could be thought of as two different types, depending on whether the relationship is contained within the one place or is between places.[3]

Causal relationships *within* a place are of three types. One type is between elements of the environment. An example that will be familiar to teachers is the effect of climate on vegetation. Green plants grow by producing complex carbohydrates through photosynthesis, and this process needs light from the sun. Plant growth also requires temperatures above 10°C and adequate moisture. The vegetation of a place consequently depends on the levels of sunlight, warmth and moisture provided by its climate. This relationship can be seen, and quite dramatically, in the vegetation on mountains in dry regions. The photos in Figures 3.3a and 3.3b show the change in vegetation on Mount Lemmon, a 'sky island' near Tucson, Arizona, in the USA. At the base of the mountain, the average annual rainfall is around 350 mm, but it is very variable and evaporation is high, and the resulting vegetation is desert scrub. Towards the top of the mountain the average annual rainfall is about 800 mm, and temperatures and evaporation are lower. Here the vegetation is pine forest. In a one-hour drive up the mountain, the vegetation changes from desert scrub through grassland, woodland and pine-oak forest to pine forest, a 1.8-km vertical transition similar to the 2400-km horizontal transition in vegetation between Arizona and Washington State. Climate therefore influences vegetation on Mount Lemmon, but it is important that students understand that this is through the mechanisms that determine plant growth.

While the effects of climate on vegetation are very evident, does vegetation have an influence on climate? This is less clear and will be examined in Chapter 10.

A second type of causal relationship within a place is that between the environment and human activities. For example, seasonally flooding rivers influenced the development of early agricultural civilisations in the Fertile Crescent in West Asia, land forms have an influence on land use, and climate has an influence on the type of agriculture, as well as on tourism and retirement settlement. People, on the other hand, have had a longstanding and major impact on the environment, and this will be examined in Chapter 6. The interrelationship between the environment and people is often the only interrelationship

Figure 3.3 (a) Sonoran desert scrub below Mount Lemmon, Arizona (elevation around 900 m); (b) Pine forest near the top of Mount Lemmon, Arizona (elevation around 2700 m).

Source: Author.

discussed in geography books for students and teachers, but it is only one of several types of interrelationships within a place.

The third type of causal relationship within a place is that between elements of human geography. For example, mining settlements, rural villages, coastal retirement places, manufacturing centres and university towns all differ to some extent in the characteristics of their populations, because of the influence of their economies on employment and the people who move into or out of each type of place. Coastal retirement places have older populations, and jobs catering for retired people, while many mining settlements now have rotating populations of workers who fly in and fly out every few weeks, and whose families live permanently elsewhere.

Causal relationships *between* places include the effects of external demand on the economy of a place, of upstream erosion on downstream sedimentation and of differences in employment opportunities and wages on migration between places. These relationships may seem quite self-evident, but they can be complex, and it is important that students clearly understand the mechanisms that produce them. While causal relationships in physical geography are autonomous and self-regulating ones determined by physical, chemical and biological laws, those involving humans are more complex, because they are the result of human decisions. Consequently, migration, for example, is not an inevitable response to economic differences between places but depends on people's assessment that their life can improve by moving, as noted earlier.

Relational thinking

Relational thinking, which has become popular in human geography, also illustrates the concept of interconnection. However, it is a difficult idea to pin down, has different interpretations and is the subject of some rather obscure writing. Tim Cresswell helpfully describes the core of the idea as follows: 'For something to be relational it has to be a product of its connections rather than a product of some essential self. … As connections (relations) change, so those things which are in relation change' (Cresswell, 2013, p. 235). A good example is Alexander Murphy's analysis of the problems of the Lake Chad Basin in Africa, outlined in the introduction to this book. He argues that:

> Simplistic views of the crisis that focus solely on population growth, ethnic conflict, or resource management practices in the Basin fail to reckon with the myriad ways in which it is deeply affected by developments originating far outside the region.
>
> (Murphy, 2018, pp. 6–7)

These include:

- Atmospheric pollution from Europe that has affected air circulation patterns and added to the effects of climate change in reducing precipitation.

- Markets and trade arrangements emanating principally from Europe that drove the expansion of water-intensive commercial agriculture.
- The influence of ideologies emanating from Southwest Asia on the rise of Boko Haram.
- Western countries' support for unpopular national governments and their actions against Boko Haram.

Some other examples of relational thinking that are relevant to school geography are:

- That the characteristics of a place are produced by its relationships with other places, such as through trade, migration and business or political decisions made elsewhere. This idea is discussed in Chapter 4.
- That economic and social space is partly produced by the personal and institutional relationships that connect places. As these connections change, so does space. This idea is discussed in Chapter 6.
- That migration is a product of the relations (interconnections) between places.
- That cities are shaped by relationships with other cities that facilitate the transfer and translation of ideas on urban policy, planning and governance between them (Jacobs, 2012).

Relational thinking could be thought of as a form of causal thinking that puts all the emphasis on the effects of external relationships and dismisses any role for internal factors. As discussed in the chapters on place and space, this may be going too far with a good idea, because it privileges one type of explanation over all others.

Holistic thinking

Holistic thinking is not a subsidiary concept of interconnection, but a description of an interconnected way of thinking; a habit of mind that looks beyond the immediate topic to its relationships with a wide range of other phenomena. It is the opposite of the reductionism common in the natural sciences, and in some of the social sciences, in which the focus is often on a very narrow set of variables and the relationships between them. In school geography, it is the ability to link separate topics together. For example, ocean currents transport heat from the equator towards the poles, which makes the atmosphere cooler in the tropics and warmer at higher latitudes than it would otherwise be. Understanding this connection between the oceans and the atmosphere is essential for an understanding of global climates, while understanding the role of the oceans in absorbing heat from rising global temperatures is also essential for forecasting the effects of greenhouse gas emissions on global warming. Studying oceanic and atmospheric circulations separately may miss these important interconnections.

Holistic thinking, or connected thinking, has regularly been advocated as a defining feature of geography. Alistair Bonnett, in his book *What is geography?*, explains that some of the pioneers of the subject, such as Humboldt, Ritter and Mackinder, 'offered connection as the core of the discipline' (Bonnett, 2008, p. 87), although the connections they were interested in were limited to those between humans and nature. Holistic thinking is often mentioned by teachers as a strength of the subject. For example, a study of the conceptions of geography held by a small group of pre-service teachers in Finland found that their views were dominated by three terms. These were spatiality, phenomena and a holistic approach (Virranmäki et al., 2019), but, like the pioneers above, their view of holistic thinking was that it was mainly about the relationships between humans and nature, and connected the natural and social sciences.

In geography, holistic thinking is not confined to human-environment relationships, but can be applied in all types of investigation. Thinking that looks beyond what can be observed to investigating underlying causal factors also can be considered holistic. For example, migration could be explained by asking migrants the reasons for their decision to move places. At a deeper level, it could be explained by identifying the factors that led to them making that decision, which could be an education system that prepared young people for work that was not available locally, lack of opportunities to earn an income, changing aspirations or the growth of economic opportunities elsewhere. At an even deeper level, the causes of migration could be the spatially uneven development of the regional, national or global economy, in turn conventionally explained by economic development or radically explained by capitalism.

The importance of holistic thinking is highlighted by Wood, who believes that in the UK 'Many students parcel their learning into convenient "bundles" of knowledge and understanding [which] can often lead to an absence of a wider appreciation of the complexity and inter-connectivity of the subject' (Wood, 2007, p. 115).

Conclusion

The concepts described in this chapter obviously overlap significantly, because what they have in common is an insistence on the importance of the interconnections between environments, people, places, organisations and events, and the limitations of viewing them in isolation, although they each emphasise different aspects of the concept. Interdependence is a particular type of interconnection; flows are the material substances and non-material information that move between places and connect them; processes are sequences of causal relationships; systems are interconnections that constitute an organised whole;

cause-and-effect interconnections are about causation; and relational thinking argues that things can only be understood through their interconnections with other things. Students may find it helpful to think of interconnection as a concept that integrates all these ideas. It has been selected as a key concept because it has many of the functions identified in Chapter 2 as powerful forms of thinking. It is a way of thinking that can be used to ask questions, organise information into systems and processes and search for explanations through causal relationships. It is also a way of perceiving the world holistically and relationally, which it could be argued is the most valuable outcome of a geographical education.

Summary

- The concept of interconnection is about recognising that phenomena do not exist in isolation but are always linked to other phenomena in complex ways. It is a high-level concept that incorporates a number of subsidiary ideas.
- Interdependence is about the mutual dependence of phenomena on each other.
- Flows are the movement of material substances and non-material information between places.
- The soil moisture budget represents the flows of water into and out of the soil and is an integrating concept in physical geography.
- Processes are a sequence of actions or events that determine the characteristics of something.
- A system is a set of interconnected entities that function together and can be conceptualised as an integrated whole. The components are linked through flows of energy and matter and, where humans are involved, people and information. Because the components of a system are interconnected, change in one produces change in others. The subsequent changes may be experienced in the same place as the initial change, and/or in different places, or at a different scale, or at a different time.
- Causal relationships are interconnections in which something is caused by something else. They can be relationships within the one place, or between places, and across time scales.
- Relational thinking contends that things can only be understood by their relationships with other things.
- Holistic thinking is a description of an interconnected way of thinking.

How could you use this chapter in teaching?

- Students could be regularly asked to look for interconnections between what they are studying and other phenomena and other places.
- Students could be regularly asked to look for interconnections between the geography topics they are studying or have studied.
- Students could be encouraged to construct what they are studying as a system, where this would be appropriate.
- Students should be identifying the mechanisms in causal relationships and understanding that association and correlation are not explanations.
- Students should be taught how to construct concept maps of relationships as a way to help them think holistically.

Notes

1. If you teach about urbanisation in 'developing' countries, Randolph and Storper (2023) is well worth reading.
2. See Lux and Budke (2020) for a discussion of the use of digital games to develop systems thinking and a list of recommended games.
3. This is similar to the distinction van der Schee makes between horizontal and vertical relations in his spatial model for the geographical analysis of a region. The model has been used in Dutch geography training programs (van der Schee, 2000).

Useful reading

Butt, G. (2017). Globalisation: A brief exploration of its challenging, contested and competing concepts. *Geography*, *102*(1), 10–17. https://doi.org/10.1080/00167487.2017.12094004

Globalisation is an example of interconnection.

Cook, S. (2018). Geographies of mobility: A brief introduction. *Geography*, *103*(3), 137–145. https://doi.org/10.1080/00167487.2018.12094050

Dessen Jankell, L., & Johansson, P. (2023). System geographical webbing as an object of knowing to analyze sustainability issues in geography. *Journal of Geography Education*, *50*(3), 119–140. https://doi.org/10.18452/25713

A review of a method to teach systems thinking through the use of a connection web.

King, R. (2012). Geography and migration studies: Retrospect and prospect. *Population, Space and Place*, *18*, 134–153. https://doi.org/10.1002/psp.685

A review of geographical studies of migration.

Rawding, C. (2013). The importance of holistic geographies. *Geography*, *98*(3), 157–159. https://doi.org/10.1080/00167487.2013.12094382

Renshaw, S., & Wood, P. (2011). Holistic understanding in geography education (HUGE) – An alternative approach to curriculum development and learning at Key Stage 3. *The Curriculum Journal*, *22*, 365–379. https://doi.org/10.1080/09585176.2011.601656

Smith, J. (2015). Geographies of interdependence. *Geography*, *100*(1), 12–19. https://doi.org/10.1080/00167487.2015.12093949

A thoughtful discussion of the various applications of the concept of interdependence.

Wood, P. (2007). Developing holistic thinking. *Teaching Geography*, *32*, 113–115. https://www.geography.org.uk/Journal-Issue/6232c61a-36d8-47c5-bc25-24835d0f4d4a

References

Bonnett, A. (2008). *What is geography?* SAGE.

Castree, N. (2015). The Anthropocene: A primer for geographers. *Geography*, *100*(2), 66–75. https://doi.org/10.1080/00167487.2015.12093958

Chapman, G. P. (2009). Systems. In R. Kitchin & N. Thrift (Eds.), *International encyclopedia of human geography* (pp. 146–150). Elsevier. https://www.sciencedirect.com/science/article/pii/B9780080449104007537

Cox, M., Elen, J., & Steegen, A. (2017). Systems thinking in geography: Can high school students do it? *International Research in Geographical and Environmental Education*, *28*(1), 37–52. https://doi.org/10.1080/10382046.2017.1386413

Creswell, T. (2011). Mobility. In J. Agnew & D. N. Livingstone (Eds.), *The SAGE handbook of geographical knowledge* (pp. 571–580). SAGE Publications. https://doi.org/10.4135/9781446201091

Cresswell, T. (2013). *Geographic thought: A critical introduction*. Wiley-Blackwell.

Fox, S. R. (2012). Urbanization as a global historical process: Theory and evidence from sub-Saharan Africa. *Population and Development Review*, *38*(2), 285–310. https://doi.org/10.1111/j.1728-4457.2012.00493.x

Jacobs, J. M. (2012). Urban geographies I: Still thinking cities relationally. *Progress in Human Geography*, *36*, 412–422. https://doi.org/10.1177/0309132511421715

Keys, P. W., Porkka, M., Erlandsson, L-W., Fetzer, I., Gleeson, T., & Gordon, L. J. (2019). Invisible water security: Moisture recycling and water resilience. *Water Security*, *8*, 100046. https://doi.org/10.1016/j.wasec.2019.100046

Legates, D. R., Mahmood, R., Levia, D. F., DeLiberty, T. L., Quiring, S. M., Houser, C., & Nelson, F. E. (2010). Soil moisture: A central and unifying theme in physical geography. *Progress in Physical Geography*, *35*, 65–86. https://doi.org/10.1177/0309133310386514

Lux, J-L., & Budke, A. (2020). Playing with complex systems? The potential to gain geographical system competence through digital gaming, *Education Sciences*, *10*(5), 130. https://doi.org/10.3390/educsci10050130

Murphy, A. B. (2018). *Geography: Why it matters*. Polity Press.

Nye, J. S. (2020). Power and interdependence with China. *The Washington Quarterly*, *43*(1), 7–21. https://doi.org/10.1080/0163660X.2020.1734303

Pooley, C. G. (2009). Mobility, history of everyday. In R. Kitchin & N. Thrift (Eds.), *International encyclopedia of human geography* (pp. 144–149). Elsevier. https://www.sciencedirect.com/science/article/pii/B9780080449104010269

Randolph, G. F., & Storper, M. (2023). Is urbanisation in the Global South fundamentally different? Comparative global urban analysis for the 21st century. *Urban Studies*, *60*(1), pp. 3–25. https://doi.org/10.1177/00420980211067926

Straussfogel, D., & von Schilling, C. (2009). Systems theory. In R. Kitchin & N. Thrift (Eds.), *International encyclopedia of human geography* (pp. 151–158). Elsevier. https://www.sciencedirect.com/science/article/pii/B9780080449104007549

Van der Schee, J. A. (2000). Helping children to analyse a changing world: Looking for patterns and relationships in space. In M. Robertson & R. Gerber (Eds.), *The child's world: Triggers for learning* (pp. 214–231). The Australian Council for Educational Research.

Vayda, A. P., McCay, B. J., & Eghenter, C. (1991). Concepts of process in social science explanations. *Philosophy of the Social Sciences*, *21*(3), 318–331. https://doi.org/10.1177/004839319102100302

Virranmäki, E., Valta-Hulkkonen, K., & Rusanen, J. (2019). Powerful knowledge and the significance of teaching geography for in-service upper secondary teachers – A case study from Northern Finland. *International Research in Geographical and Environmental Education*, *28*, 103–117. https://doi.org/10.1080/10382046.2018.1561637

Wetzel, R. G. (2001). *Limnology: Lake and river ecosystems* (3rd ed.). Academic Press.

Wohlleben, P. (2015). *The hidden life of trees: What they feel, how they communicate – Discoveries from a secret world*. Black.

Chapter 4

Thinking geographically

Place

Place as a concept is about how we think about, perceive, experience and explain places; how they are the geographical context in which things exist and events happen; how we can understand their influence and why they matter; and how we can use place to think geographically. It should not be confused with the study of places as objects.

Conceptualising place

What is a place, and how can it be defined? Tim Creswell (2014, p. 4) writes that 'Place is best known to geographers as a meaningful segment of space'. This means that it is an area of the Earth's surface, a space, that has been identified and named by people. A place has the following characteristics:

1 A location.
2 A spatial extent, but this may be quite fuzzy and undefined, and places merge into each other. Many do not have clear boundaries, but they are still bounded and not limitless.
3 A name, identity and meanings for individuals and social groups. These identities and meanings are created by people and are social constructions, and consequently the names, identities and spatial extent of places can change over time. Places can also change their name, identity and meanings when national boundaries shift, as in Europe after World War II, or through colonisation, as in the Americas, Australia and New Zealand, where Indigenous place names were replaced with European ones, and in some cases are now being returned.

Places defined in this way are described by Creswell (2014) as unique assemblages of materialities (i.e., material things), meanings and practices. Materialities include the biophysical environment of landforms and vegetation and the built environment of buildings, roads and other structures. Meanings include names, histories, stories and memories. Practices include the everyday activities of living, working, learning, recreating and simply passing the time. Each

DOI: 10.4324/9781003376668-5

place has its own combination of these attributes, and this makes each place distinctive. Places identified by these criteria range in size from a room to a home, a suburb, town or city, region (see Box 4.1), nation and even the whole planet, because these are all spaces that have been identified and given meaning by people. However, the places of interest to geographers, and studied in schools, are likely to be in the middle of this range, such as suburbs, towns, cities and regions. Comparative studies of countries may also be thought of as being about places.

Box 4.1 Region and territory

A region is a type of place, because it is an area of space that has been defined and given meaning by people. Regions have historically been identified as belonging to two main types:

1. Homogeneous, formal or uniform regions, which consist of areas that are similar in one or more of their characteristics. They could be climatic regions (areas with similar climates), biomes (areas with similar biotic environments), economic regions (areas with the same economic base, such as manufacturing or tourism), administrative regions (such as the States and Territories in Australia, counties in the USA and local government areas in the UK) and social regions (areas with a common history, ethnicity, culture and identity, such as Catalonia in Spain or the Bible Belt in the USA). Wine and food regions, especially in Europe, are another example of this type of region. The French term 'terroir' describes places with distinctive landscapes, soils, climate, products and rural culture that are famous for their production of particular types of wine and cheese. Their significance is recognised by their designation as legally protected areas which have been precisely mapped, known as Protected Designations of Origin in the European Union (Hill, 2019).
2. Functional regions, which are areas that are integrated through the functional connections between the parts of the region, such as the area for which a city or town is the marketing, processing, transport or service centre. A water catchment might also be described as a functional region, as it is an area that is integrated by flows of water.

However, regions can also be areas that people, government agencies or businesses have identified for a particular purpose, which could be economic development, addressing social disadvantage, managing the biophysical environment or operating a business. A region can be of any size that suits the purpose for which it is being defined, ranging from groups of suburbs to groups of nations, or from a local vegetation association to a global biome. All are human constructs, as there is no objective set of regions out in the world waiting to be discovered.

Territory is another place concept and is defined as 'A unit of contiguous space that is used, organized and managed by a social group, individual person or institution to restrict and control access to people and places' (Gregory et al., 2009, p. 746). The term is typically associated with national or regional governments.

Different people may perceive the same place differently. For example, fifth-generation residents will have quite different perceptions and meanings of their place compared with recently arrived migrants, and teenagers are likely to have different (and possibly less positive) perceptions compared to their parents. The consequences of this view of place are succinctly described by Denise Freeman and Alun Morgan:

> These alternative approaches also acknowledge that there is not necessarily one true reading of place but rather a diversity of idiosyncratic individual and group-based understandings, some powerful, some marginalised. This gives rise to an understanding of place that acknowledges and seeks to include a diversity of people and voices, each with their own opinions and perspectives on places, and with the power and agency to shape those places—either deliberately, or through the practices they engage in, in their everyday lives.
>
> (Freeman and Morgan, 2017, p. 121)

Creswell (2015, p. 42) adds: 'place does not have meanings that are natural and obvious but ones that are created by some people with more power than others to define what is and is not appropriate'. However, the extent of this diversity of meanings and perceptions varies, with places within large global cities being much more diverse than small villages, and we should be careful not to generalise from examples of particular types of places, or from Western ways of thinking about places.

Because of their different characteristics, identities and power, people use and experience places differently. Physical disabilities limit some people's access to or use of particular places, while gender, ethnicity or sexuality may make others avoid places they perceive as hostile or unsafe (Hille, 1999). Some may be excluded from particular places, such as homeless people, skateboarders or people with a different ethnicity to the dominant group. Young people may feel unwelcome in some places, while in cities with predominantly European populations women wearing Muslim headdress may be abused and threatened. In many countries, women's use of public places is either controlled or discouraged, as in Iran. These exclusions from places and spaces can be and are contested and changed, for example by women's movements such as Reclaim the Night and Reclaim the Streets or by regulations to make public space and buildings accessible to people in wheelchairs.

Some public places have symbolic power and have been used by protesters to draw attention to their cause because they are associated with governmental or corporate power, such as the Red Square in Moscow and Wall Street in New York. For example, in Egypt during the Arab Spring, crowds gathered in Tahrir Square in 2011 and again in 2013 to demonstrate against successive Presidents. In China, Tiananmen Square in Beijing has been the site of several major events in Chinese history, including student-led protests in 1989. Its symbolic power is such that any commemoration in China of the events in the Square in 1989 is prohibited. Other places have religious power, such as Rome for Catholics, Banaras for Hindus, Mecca for Muslims and Jerusalem for Jews.

In physical geography, place has a somewhat different meaning to that described above but still similar in some important ways. Places in physical geography are areas at all scales that are identified as having some common characteristics. For a study of a river, places might range from an erosional or depositional landform (a geomorphic unit), to a section of the river with an assemblage of landforms (a reach), to a complete river basin. For vegetation, places could be defined as areas of distinctive plant communities and their edaphic, geomorphic and climatic environments (Figure 4.1). These are not places identified by their meanings to people, and they may not be named, but

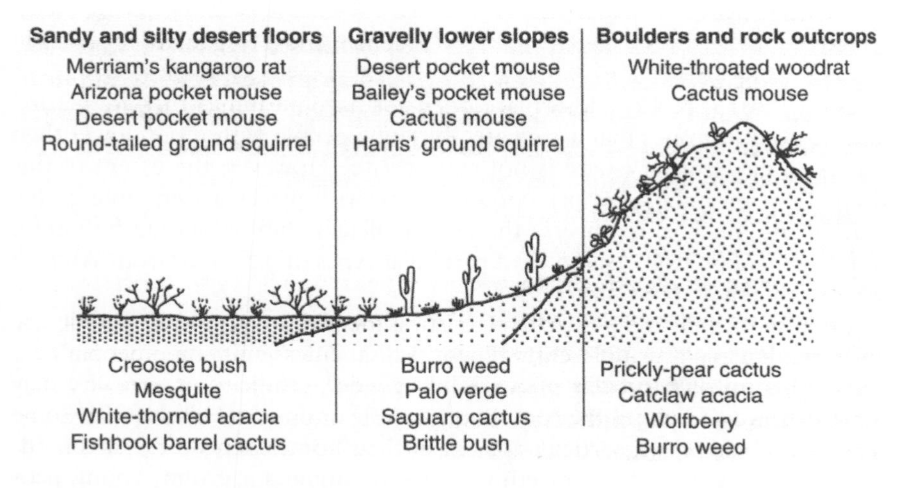

Figure 4.1 Types of place in North American deserts.

Source: Gregory (2009), p. 182. Copyright SAGE.

they are places with unique environmental characteristics, and these have an influence on environmental processes.

Indigenous concepts of place

Indigenous concepts of place are different to those described above, and can be illustrated from Australian Aboriginal culture. In Aboriginal Australian English, the term Country refers to an area or place known to be associated with and belonging to a group of people linked by kinship, although because of dispossession and migration they may no longer live there. Country is therefore a type of place. In other Indigenous cultures, the equivalent English term may be Land. Country is eloquently described by Ambelin Kwaymullina, an Aboriginal lawyer and writer, in this passage:

> For Aboriginal peoples, country is much more than a place. Rock, tree, river, hill, animal, human—all were formed of the same substance by the Ancestors who continue to live in land, water, sky. Country is filled with relations speaking language and following Law, no matter whether the shape of that relation is human, rock, crow, wattle. Country is loved, needed, and cared for, and country loves, needs, and cares for her peoples in turn. Country is family, culture, identity. Country is self.
>
> (Kwaymullina, 2005, p. 12)

For Aboriginal Australians, the concept of Country as a place has these characteristics:

- Country was created and given meaning by Ancestral Beings whose knowledge is passed on between generations.
- A Country is small enough to accommodate face-to-face groups of people and large enough to provide the resources to sustain them.
- A Country is 'politically autonomous in respect of other, structurally equivalent, countries and at the same time is interdependent with other countries' (Rose, 2005, p. 295).
- Country contains the stories, songlines and sacred sites that preserve the knowledge and laws of the community, so Country is a teacher.
- Country is a living whole. Weather, land, sea, plants, animals and people are all interconnected and interrelated, both practically and spiritually. There is no concept of nature and culture as separate.
- Country has power, and traditional owners must obey the protocols of announcing their arrival and asking to be recognised and safe while they travel through Country.
- Country is aware, and can communicate. Some non-Indigenous students may find this difficult to understand, but the concept could be illustrated by examples of this communication. These include birds that warn people

of danger, fireflies that tell when goose eggs and crocodile eggs are ready to harvest, flowers that tell when to harvest a plant or hunt an animal, birds that call when the tide is coming in and winds that tell when the conditions are right for fishing. People must also be open to what Country is telling them, for example, that it is sick and that degraded areas need attention, or the appropriate ceremonies are needed to restore its health.
- The health of Country and people are interdependent.

This is a much richer concept of place, and of relationships between places and humans, than the conceptualisation described at the beginning of the chapter. It is also a way of thinking that produces strong place attachment and strong political movements to maintain or restore Indigenous people's rights to their historic territories. Indigenous concepts of place such as these remind us not to assume that Western geographical ways of thinking are universal.

Explaining what places are like

Explaining why places are like they are is an integral part of the geography curriculum in many countries. What places are like can be described by their characteristics, which include people, climate, economic activities, landforms, buildings, infrastructure, soils, vegetation, communities, water resources, cultures, events, governments, services, mineral resources and scenery. Some are produced by the environment of a place—its climate, landforms, soils, vegetation and natural resources—as these not only determine its environmental characteristics but also influence (but do not determine) its built environment and economic activities. For example, the environment of a place may support intensive agriculture (as in areas watered by seasonally flooding rivers); the development of ports (as in places with natural harbours accessible from sea and land); particular types of house construction (depending on the availability of timber, clay or stone); or the growth of industries based on local resources.

However, most of the characteristics of a place have been produced by people, acting as individuals, businesses, organisations and governments, who have cleared vegetation, modified rivers, established ports and created built environments, industries, communities, forms of governance and cultures. It is important for students to understand that places are made by people through human agency (i.e., people and groups taking actions), because this may give them the idea of being involved in shaping their own place. Many of these agents are local people, businesses and government, but others are located outside the place. Some places are quite new creations, such as mining towns and national capitals (e.g., Canberra in Australia and Brasilia in Brazil) established and developed by corporations and governments. It is also important for students to understand that change can be the subject of conflict between different interest groups with different degrees of power and that the ability of local communities to control change is constrained by the power of governments, businesses and economic interests and forces beyond their control.

Many of the influences on places are external, because places are strongly affected by their economic, demographic, cultural and political relationships, or interconnections, with other places. Examples of the effects of these external relationships include:

- Loss of retailing in a small town because of competition from larger centres.
- Growth of a new industry created by the investment decision of a non-local corporation.
- Loss of employment through the decisions of non-local corporations to close a local business.
- Changes in external demand for goods or services exported from the place.
- Changes in the composition of the population from internal or international migration.
- Cultural change produced by fashions, practices and ideas from elsewhere or by migrants from within or outside the country.
- Decisions of a national government to centralise or decentralise government functions, leading to a loss or gain in employment and local control.
- Neglect by governments whose political interests are focused elsewhere, or whose ideology leads to economic restructuring that disadvantages some places.

Note that these external relationships are not really between places, but between people, organisations, businesses and governments in these places. Places have no causal powers of their own; for example, one place or country cannot exploit another place or country, because it is 'particular classes, institutions or individuals in one area which exploit those in another area' (Sayer, 2000, p. 112).

This emphasis on external relationships is a relational view of place, which contends that they are a product of their relationships with other places and that much of what seems intrinsic to the character of a place is the product of past relationships. Freeman and Morgan describe a relational idea of place as:

> Places are no longer seen as spaces with inherent characteristics, bounded and discrete, but as the outcome of processes and forces operating over, across and between them; forming an interdependent web of relations. Such a perspective is more likely to see places as 'nodes' in a series of complex networks operating across various types of spaces (economic, political, environmental, etc.)
>
> (Freeman and Morgan, 2017, p. 121)

John Agnew (2005) explains that a relational view of place is part of a continuing debate about the relative importance of place and space. Some geographers consider place to be a concept about the past and associated with localism, parochialism and reactionary views. Space, on the other hand, is modern, progressive and radical, and the appropriate arena for political action. Place is associated with humanistic geography, while space is associated with both spatial science and Marxist geography, although for very different reasons. This view of place has been criticised on several grounds.

- It is largely based on the example of London and the southeast of England, and there are many places in the world that are much less globally connected and much less diverse. It does not describe Indigenous perceptions of place.
- The concept of places as nodes is difficult to reconcile with the reality of people's attachments to and feelings about places that to them are areas rather than nodes, and real entities.
- It is unable to explain the existence of regional movements for greater autonomy and even independence, which are based on people's sense of attachment to a discrete and distinctive area.
- It dismisses the role of place as an arena for local action. Ash Amin, for example, argues that because any 'particular geographical site can only ever be a nodal connection in a hydra-like network space that never coheres into a local public sphere' (Amin, 2004, p. 38), the scope for local and regional action is limited. Others, while agreeing with the significance of extra-local relationships, point to the continuing use of regions to design and deliver economic development and welfare programs, and the roles of place-based agencies (Jones, 2010). Place-based public policies are discussed later in this chapter and also in Chapter 11.

Cresswell concludes that: 'It makes no sense to dissolve place entirely into a set of flows and connections. At any point in time, a place is a particular combination of materialities, meanings and practices that encourages some connections and makes others unlikely' (Cresswell, 2014, p. 20). The debate continues, and what it illustrates is that human geographers, like other social scientists, sometimes pursue new ideas to the unhelpful exclusion of older ones.

A final way of thinking about how to explain why places are like they are is the idea that 'Places are created things and tend to reflect or mediate the society that produces them' (Cresswell, 2008, p. 136). This viewpoint is a product of Marxist and other critical thinking that emphasises the influence of the structure of a society and its economy on its social and economic geography. For example, the more unequal a society the greater is likely to be the social and economic differences between places, differences which may be compounded by racial discrimination. Similarly, the more that political and economic power is decentralised the more opportunities there are for small places to survive. As an example, Richard Walker argues that in the USA:

> an extreme form of decentralized sovereignty [and] the close alliance of capital and the state at the local level … has allowed for open competition among places, compensation for the capitalist bias towards big cities, and the flowering of a highly dispersed system of cities and regions, making the US the most decentralized of modern economies.
>
> (Walker, 2020, p. 546)

In this the USA contrasts with Australia, where political and economic power (and consequently employment and population) is centralised in the capital cities of the states, and there are few medium-sized cities, as discussed in Chapter 7.

To conclude, to understand why a place is like it is we should look at:

- Explanations from within the place, such as the influence of its relative location (explained in the next chapter), environment, history, culture, infrastructure, buildings, governance and the actions of its people.
- Explanations from its relationships with other places.
- The social and economic structure of the region or nation within which a place is located.
- How these interact with each other and with the existing characteristics of the place.

It is also important for students to understand that places are constantly changing in their characteristics, and that even when there are few visible signs of changes in the built environment of a place, there may be changes in the people who live there, and in what they do.

Attachment to place

People develop attachments to particular places by living in and becoming familiar with them, and this place attachment is significant in that it contributes to a person's identity and sense of belonging. It also contributes to the emotional development of children, which makes it a concept of relevance to teachers. Sarah Little and Tori Derr (2018, p. 15) write that 'Much like with human attachment, children gain a sense of their self-worth and self-identity from attachment to place'. The foundations for this attachment are formed in middle childhood, during the primary school years, and geography has a role in this. A geography that teaches students about their own place and what it is like, how it supports their lives and how they are connected to it and to the people who live in it, can help to develop this sense of belonging and attachment and is an important but largely neglected contribution of school geography to the development of children.

Place attachment is likely to be strong when a person's close relationships are with people in the same place and where they belong to a strongly functioning local community. Peter Smailes (2000, p. 159), building on several decades of research on rural communities in South Australia, writes of 'the persistent, deep-seated importance of local, personal interaction with real people in familiar place settings in the formation of identity and a feeling of place-belonging'. Place attachment is also likely to be strong amongst people who do not move far from their place of origin, which is the case with the vast majority of the world's population in both high- and low-income countries. It is especially strong amongst many Indigenous communities, even when they have been displaced or moved away from their original home place. On the other hand, place attachment has often thought to have declined for the growing number of mobile people, yet research suggests that this may not be the case, because mobile people develop attachments to new places without necessarily losing their attachment to their place of origin and can become engaged with many places. Place attachment is not incompatible with mobility.

The significance of place attachment is further demonstrated in the sense of loss experienced by people displaced from their place of origin, such as refugees and Indigenous peoples. Jeff Malpas further notes that the destruction of places is:

> currently being used in many places of the world as a means to destroy or undermine communities and cultures through the destruction of the material basis of those communities and cultures, including the destruction of buildings and sites, or restriction of access to them. The embeddedness of culture in place, and the destruction or control of place as one of the means by which authority is imposed and maintained is a key reason for the intractability of the political situation in Israel and Palestine.
>
> (Malpas, 2018, p. 199)

These examples of the significance of place attachment provide further evidence that a place is more than a 'nodal connection in a hydra-like network space'.

Not all feelings about places are positive. Home, for example, may be a safe haven or a place of danger for women and children. Neighbourhoods may be threatening to minority groups, and towns may be hostile to people who do not share the dominant values and do not 'fit in'. Place attachment can also be negative. Inward-looking communities, with strong place attachment, 'can be stifling, hide bound in tradition, inbred, and, especially for the young, boring, and dull. It is not surprising that getting out of a place is a recurrent theme in popular music and movies' (Relph, 2017, p. 182).

Much of the research on place attachment has been by psychologists and sociologists, but geographers are also involved and contribute a deeper understanding of the distinctive characteristics of places, and how these influence place attachment.

Using place to explain

Places are not just something to be conceptualised, interpreted and explained; they can also have an influence and be an explanation. This is because places are both the primary field of human experience—'where people live and move and have their being' (Casey, 2018, p. ix) and as 'milieux that exercise a mediating role on physical, social and economic processes and thus affect how such processes operate' (Agnew, 2011, p. 317). In other words, places affect us because they are the geographical context in which we live, and they affect processes and events because they are the geographical context in which they happen. As Paul Adams writes:

> Geographical interest in place stems from an assumption that processes of all sorts (political, economic, cultural, biological, geological, and so forth) happen *with* places rather than merely *in* places. This distinction is subtle, but important: for geographers, processes are shaped, affected, reworked, modulated, organized, coordinated, and inflected by places.
>
> (Adams, 2017, p. 2)

What this means is that places are where processes interact with the unique characteristics of a place to produce different outcomes in different places. For example, changes in the structure of an economy produced by national and global economic influences will disadvantage some places but advantage others, depending on the characteristics of their individual economies. Some may lose employment because of competition from imported products that they previously made, while others may gain from new export opportunities for their industries. Similarly, the impact of a tsunami on a coastal region varies from place to place according to the environmental characteristics of each place. Coasts where the topography channels the tsunami wave into a smaller space will increase its impact, whereas coastal mangrove forests will reduce the height and speed of the waves. This is why Joseph Holden says that the value of case studies in physical geography 'is because what causes change is the interaction of mechanisms with particular places' (Holden, 2011, p. 13).

Another consequence of the uniqueness of places is that people in different places may respond differently to the same issues and events. This is partly because different places have different population compositions, as measured by ethnicity, age, education and income. Once the effects of these variables are removed, what is left reflects the contextual role of place. For example, in the UK 2016 referendum on whether to leave the European Union, there was a major difference between inner London (the cosmopolitan core of the country) and the rest of England that could not be statistically accounted for by differences in the characteristics of their populations (Johnston et al., 2018). Views on continuing membership of the EU were clearly different in these two parts of England, as well as in Scotland and Northern Ireland, although the reasons for this difference continue to be debated.

For the same reasons, similar problems may require different strategies in different places, as what will work in one place may not work in another. Alexander Murphy provides this example:

> An effort to promote economic development by offering tax breaks to lure business investment may have positive socio-economic benefits in one place, but negative ones in another depending on the mix of existing businesses in nearby communities, local employment opportunities and labor conditions, and the attitudes of the local community.
>
> (Murphy, 2018, p. 86)

The influence of place on people's lives

The following sub-sections discuss two examples of the influence of place, through the characteristics of places, on our lives and wellbeing.

Health

> Scottish men live more than two years less than English men, and Northerners in England live two years less than Southerners. Londoners living in Canning Town at one end of the Jubilee tube line live seven years less than those living eight stops along in Westminster. There is a 25-year gap in life expectancy between residents of the Iberville and Naverre suburbs of the US city of New Orleans, although they are just 3 miles apart.
>
> (Bambra, 2017, p. 1)

The places in which people grow up and live can have an influence, although not a determining one, on their health, often very simply measured by life expectancy. There are major differences in health between places in all countries, particularly at the scale of small areas. Research to understand these differences, and the role of place in them, forms the thriving interdisciplinary field of health geography, in which geographers, epidemiologists, public health researchers and other social scientists are all involved.

An example of the influence of the characteristics of places on health is a study of declines in heart disease mortality in the US South from 1968 to 2014. The decline varied between counties, and the variations for Blacks (but not for Whites) were statistically associated with historical differences in the concentration of slave labour, and its continuing legacy of inequality and racial discrimination. The researchers conclude that:

> Between 1968 and 2014, Blacks in counties with the highest 1860 concentration of slavery experienced slower rates of mortality declines compared

> to Blacks in counties with a history of lower slave concentration. Nearly half of this association is explained by inter-county 20th century differences in educational and economic racial inequalities, and the cumulative use of lynchings to enforce Black subordination.
>
> (Kramer et al., 2017, p. 615)

Health geographers explain health differences between places in several ways. One is a compositional approach that attributes poor health to the life styles (such as levels of smoking, alcohol consumption, diet and physical activity) and socioeconomic characteristics (such as income, education and occupation) of the people who live in a place. This argues that poor people produce poor health. A second approach is contextual, and attributes poor health to the social, economic, and physical characteristics of a place. It argues that poor places produce poor health. The characteristics of places that can affect health include:

- The types of employment available.
- The availability of affordable fresh food.
- Opportunities to exercise.
- The provision of health and other services.
- Land, water and air pollution.
- Access to green spaces.
- The strength of social networks within the community.
- Social inequality.
- The perceived reputation of a place.[3]

These two approaches are complementary and interconnected, in that 'the health effects of individual deprivation, such as lower socioeconomic status, can be amplified by area deprivation', or reduced by living in a place with more resources (Bambra, 2018, p. 1). Clare Bambra, however, goes a further step in explanation by adding structural and political factors, as in this explanation of the causes of stroke or heart disease:

> The immediate clinical cause could be hypertension (high blood pressure). The proximal cause of the hypertension itself could be compositional lifestyle factors, such as poor diet, of which the contextual cause might be living in a low-income neighborhood. The causes of the latter are political—low-income neighborhoods exist because the political and economic system allows them to exist.
>
> (Bambra, 2018, p. 32)

This is an example of explanation at different levels, an approach described in Chapter 2.

Education

The effect of place on educational opportunities and attainment is a controversial and complex area of research. It is also difficult to synthesise, because studies use different measures of opportunity and attainment, and different statistical methods. The results of a selection of research projects are outlined below.

1 Neighbourhood socioeconomic status (SES) may influence the type of secondary education a child can access. In the Netherlands, children are allocated to one of several levels of secondary education at the end of primary school, and the most important factor in this allocation is the 'advice' from the child's primary school teacher. A study that classified children according to the SES of the neighbourhood of their school found that, after controlling for other factors:

> children living in a very high SES neighbourhood are significantly more likely to receive advice for a higher educational level than for basic/advanced pre-vocational education compared with children living in a medium SES neighbourhood (Odds Ratio, OR: 1.5). Conversely, living in a very low or low SES neighbourhood significantly decreases the odds of receiving higher advice, with respectively 28% and 22%, compared with living in a medium SES neighbourhood.
>
> (Kuyvenhoven and Boterman, 2021, p. 2671)

2 Residential location may limit a child's ability to attend schools that are perceived as higher performing. A study of student access to schools in East London (UK) found that:

> All the East London boroughs we have examined in detail in this paper have a clear hierarchy of schools in terms of popularity. Given that the number of applicants at the most popular schools exceeds places by a considerable margin, they all operate what is effectively a system of rationing places using the distance to school criteria.
>
> (Hamnett and Butler, 2011, p. 492)

The researchers go on to point out one of the consequences of this system:

> This in turn feeds through into the housing market pushing up house prices and rental levels in the streets around such schools. As a consequence, the operation of distance-based rationing in the supply of school places in the face of inelastic demand for such places serves to reinforce and reflect existing patterns of residential social segregation.
>
> (Hamnett and Butler, 2011, p. 493)

In Australia, and no doubt other countries, real estate agencies frequently identify the school zone in which a property for sale is located where this will be a selling feature.

3 In cities where different socioeconomic groups tend to be spatially segregated into separate areas, disadvantaged neighbourhoods are likely to have lower educational outcomes than the more advantaged. For example, an Australian study that mapped primary school scores on a standard test of literacy and numeracy found that, in the major cities, few schools in advantaged suburbs ranked below average, while in disadvantaged suburbs few schools ranked above average (Smith et al., 2019). In other words, the socioeconomic composition of the school had an influence on educational outcomes additional to that of the socioeconomic background of individual students. Other research suggests that disadvantaged suburbs may have an influence on children's educational attainment through:

   - Low aspirations resulting from a history of poor employment.
   - Poor self-esteem because of the perceived reputation of the neighbourhood.
   - Lack of participation in out-of-school structured activities.
   - Social isolation from people in neighbouring more affluent suburbs.
   - Lack of access to opportunity structures beyond their own suburb.
   - Limited educational resources such as libraries and computer facilities.
   - The quality of the schools.

Quantitative measures of the influence of these factors from a number of countries suggest that this ranges from small to moderate, but qualitative studies point to a significant effect on a minority of children.

Critics of the contention that disadvantaged suburbs influence educational attainment do not necessarily disagree that this influence exists but argue that it misses the fundamental causes of the problem. Tom Slater (2013) argues that the statement 'where you live affects your life chances' should be inverted to 'your life chances affect where you live'. His point is that income inequalities and the private housing market combine to determine where low-income people are forced to live. His argument is statistically supported by research on eight European cities, which found that increases in national income inequality led to an increase in socioeconomic segregation ten years later, as housing markets adjusted to the changes in income distribution. The researchers argued that increasing residential segregation was mostly driven by the highest income groups, as they had the most freedom to choose where to live and were also able to displace low-income groups from inner-city locations (Tammaru et al., 2020). In some countries, residential segregation is also caused by racial discrimination, which severely limits the ability of particular racial groups to choose where to live (Pearcy, 2020; Smith, 2001). These examples illustrate the ways that the structures of a society influence its spatial patterns, which in turn reinforce the structures.

This part of the chapter concludes with two overviews of the influence of place on people's lives. The first is by Susan Smith, a British geographer, who writes:

> Residential space is a gateway to services, educational, and employment opportunities, health, and social services. Where people live is therefore a reflection of who they are, but it is also an influence on who they can become. It has therefore been argued that tackling the inequalities embedded in residential segregation is a key priority for public policy and an urgent focus for social concern.
>
> (Smith, 2001, p. 548)

The second is by Jeff Malpas, an Australian philosopher who has written extensively about place:

> Put simply, what we are depends on what we can do, and what we can do depends on where we are situated. It is not merely, then, that we look to the places in which we live as that by means of which we explicitly articulate a sense of ourselves, but more than this, the very shape of our lives is determined, implicitly and explicitly, by the possibilities that are given in and through the places [in] which we live and our interaction with those possibilities and places.
>
> (Malpas, 2006, p. 2)

Using place to organise data

Data that are classified by places, which could be settlements, local government areas, electoral districts, regions or countries, are examples of the common use of place to organise information, whether to present it as data or to prepare it for analysis.

Using place to analyse

'Places are natural laboratories for the study of complex relationships among processes and phenomena' (Rediscovering Geography Committee, 1997, p. 30). Geographers usually cannot conduct experiments to identify cause-and-effect relationships, except in some areas of physical geography, but they can make controlled comparisons of places to test relationships between selected variables. This is an idea that goes back at least to the distinguished American geographer Richard Hartshorne, who wrote:

> the ways in which separate areas [i.e., places] are alike is no less significant than the ways in which they differ. Comparative study of such areas permits

> geography to approach the methods of laboratory sciences, in which certain facts are controlled as constants, while other vary.
>
> (Hartshorne, 1960, pp. 15–16)

An example of this type of analysis comes from a study of whether globalisation has made cities similar. If globalisation is the dominant influence it is often portrayed to be, then this is a plausible expectation. However, it conflicts with the geographer's contention that places are different, and that consequently the outcomes of the same process will be different in different places. To test the effects of globalisation Ronald Horvath compared the inner areas of Los Angeles and Sydney, two cities which are comparable in their global status. He found that instead of becoming similar through globalisation, there were major differences between them in their built environment, social geography and patterns of disorder, and that these differences had increased over time. For example:

> Los Angeles has extensive slums, residential disinvestment, nonresidential land use invading residential neighborhoods, and very little conversion of brownfields into luxury apartments. Inner Sydney, in sharp contrast, has no slums, massive investment in the residential built environment, residential land uses replacing nonresidential properties, and brownfields being generally redeveloped into luxury apartments. Furthermore, Sydney is considerably less automobile dependent than Los Angeles.
>
> (Horvath, 2004, pp. 102–103)

Horvath concludes that the explanation for these differences between the two cities is that they are different places, with different populations, different social relations and different placemaking practices and, as a result, globalisation did not produce similar cities. 'Place' makes a difference. Karen Lai also argues for the importance of studying global cities as places, and not just as nodes in networks of flows, because as places, global cities have distinctive characteristics which influence the ways each one is integrated into the global city network, and the particular roles that each performs. She writes that: 'viewing global cities as places means recognising that they are inhabited by real people and communities whose social and political lives shape, and are in turn influenced by, processes of global city formation' (Lai, 2009, p. 1009).

The methodology of comparative study that Horvath's study illustrates is described below, based on and adapted from Wilke and Budke's (2019) review of comparison as a method in geographical education in secondary schools:

1 Develop a geographical question that can be answered by a comparison of cases. A geographical question is one based on one of geography's major concepts, and in the example the concept is interconnection, which underlies globalisation.

2 Choose cases that can produce an answer to the question when they are compared. In the example, the two cities are approximately equal in their global status, so can test whether globalisation produces similar outcomes.
3 Identify the characteristics to be used to identify whether the expected outcomes are similar. In the example, these are the built environment, social geography and patterns of disorder, all of which can be described either quantitatively or qualitatively.
4 Identify other characteristics of the cases that could also affect the outcomes, as independent variables that could be causal factors in addition to globalisation. In the case study, these are the composition of the population, the nature of the societies and placemaking practices.
5 Identify the similarities and differences between the cases. In the example, despite similar exposure to globalisation, the outcomes, as measured by the built environment, social geography and patterns of disorder, were not similar. This is explained by the independent variables analysed in the study.
6 Formulate an answer and reflect on the possibility of alternative answers.

In schools this comparative method could be used to investigate:

- The influence of climate on ways of life, by selecting several places in the world with a similar climate and finding out whether the ways of life in these places were similar or different. Independent variables that might also affect ways of life could be incomes, technology and culture.
- Why some settlements are growing while others are declining, using information on population growth (the dependent variable), and possible causal factors (independent variables) such as population size, type of economy and relative location.
- The different impacts of storms on wave-dominated and tide-dominated coastlines (Brooks, 2017).

Generalising and applying

The uniqueness of places does not mean that they can only be understood as singular phenomena and that generalisation in impossible. Places share similar characteristics, such as climate, topography, relative location, type of economy and population size. Comparisons can then be made to see if a common characteristic, such as being in a remote location or being based on mining, results in other common characteristics, such as the composition of their populations or their political affiliations. For example, a study of the redevelopment of an old port district, such as the London Docklands, or an old inner city area, such as in New York, might result in the following generalisation:

> The redevelopment of old urban areas by private capital is likely to lead to the displacement of existing disadvantaged social groups, because they may

be priced out of the area by rising housing costs and do not have the skills to gain employment in the new industries being established.

This generalisation could then be applied to other urban redevelopments to see if they have experienced the same displacement outcome. If any have not, how can this be explained? The answer might be found in the governance arrangements for the redevelopment in which communities and local government have more power over the design of the project (Jones, 2017).

Place and public policy

Place is not just an academic idea but is the framework for a wide range of policies and initiatives designed to improve social, economic and environmental conditions by recognising the uniqueness of individual places. Most national economic and social policies are spatially blind in that they do not vary from place to place or target particular places. Place-based policies, on the other hand, 'target the specific circumstances of a place and engage the community and a broad range of local organisations from different sectors as active participants in their development and implementation' (Government of Victoria, 2020). Their distinctive features are that they are tailored to the needs and potential of each place and depend on local leadership and local involvement (Beer et al., 2020). They address a wide range of issues, such as educational quality and school attendance, health, disengaged youth, crime, employment, environmental quality and infrastructure, and the areas they cover range in size from a group of suburbs to a local government area to a large region. Chapter 11, on Unequal places, examines why some places are economically and socially disadvantaged compared with other places within the same country or the same city, and the policies that can be implemented to reduce their disadvantage.

Places are also where people can learn to form a society. Massey explains this idea:

> A number of geographers are now trying to use our work on place to get some messages across. ... What we want to emphasise is a notion of place as one of the arenas where people (of all ages) learn to negotiate with others—to learn to form this thing called society. It is a practice of daily negotiation which we could understand as the beginnings of democracy. In a way it is incredible that places—most places—'work' as well as they do. And when they break down we should not try to force upon them an old notion of coherence. Because a healthy democracy requires, not pacification into conformity, but an open recognition of difference, and an ability to negotiate it with mutual respect.
>
> (Massey, 2002, p. 294)

This is a powerful idea that students could be encouraged to think about. Malpas (2012) adds three more areas in which place could be a basis for effective societal responses to current concerns. One is the role of a strong sense of place in community action on environmental sustainability. A second is that in countries colonised by Europeans, reconciliation with the indigenous peoples requires a better understanding by the colonisers of the significance of place to First Nations people, especially in Australia. The third is the importance of place attachment in socially just urban development and regeneration programs, which links back to Massey's thoughts above on places as arenas where people learn to work together.

Conclusion

Places are fundamental to human existence, because we are always in a place and are consequently always influenced by a place. As a way of thinking geographically, the concept of place means being aware of the influence of people on places and of places on people and processes. It explains the wonderful variety of the world, counters the supposed universalising influence of processes such as globalisation and guards against excessive generalisation. As geographers say, 'place matters'. It also has political significance, as Ron Martin (2001, p. 203) argues that a primary objective [of geographers] 'must be to demonstrate the crucial "difference that place makes" in the construction, implementation and impact of public policy'. The ideas about place discussed in this chapter also teach students new ways of thinking about the world and about themselves, as well as ways of explaining and analysing, and they give students the knowledge that enables them to understand their own place and to participate in debates about its future.

Summary

- A place is a space that has been identified and given meaning by people, but its identity and meaning may differ between the different groups and individuals within a place.
- Places are unique assemblages of materialities, meanings and practices.
- People perceive and experience places differently.
- Places can have symbolic power.
- Regions and territories are types of places.
- Places in physical geography are areas at all scales that are identified as having some common characteristics.
- Indigenous cultures have a different concept of place, in which places precede people instead of being identified by them, are a living whole and can communicate, teach and have agency.
- The characteristics of a place are produced by internal and external influences and the interactions between them, but most are

produced by people acting as individuals, businesses, organisations and governments.

- Places are constantly changing.
- People's sense of attachment or belonging to a place or places is often a part of their identity.
- Each place is unique in its environmental and human characteristics, including its relationships with other places, and this uniqueness influences the local outcomes of environmental and human processes.
- Places influence people's lives in a wide variety of ways.
- Place can be used to organise data.
- Causal relationships can be investigated through a controlled comparison of the characteristics of places.
- A generalisation based on a study of one place can be applied to the understanding of other places.
- Place provides a conceptual framework for a range of social, economic and environmental initiatives.

How could you use this chapter in teaching?

Questions based on the content of this chapter can be grouped into three types. They start with questions that are suitable for primary school students and end with ones appropriate for upper secondary school.

Understanding my place

- What is this place like? How is it different to other places I know?
- Why is it like it is?
- What makes it a place?
- What is the composition of the population of this place, how diverse is it and how is it changing through natural increase and net migration?
- If your place has an Indigenous population, how do they describe and perceive the place?
- Are there particular parts of your place that are identified with particular groups of people?
- Are some people unwelcome or excluded from this place?
- How do marginalised people live in and use the place? (see Bustin, 2011a, 2011b, 2019)
- How do young people use spaces in this place?

- How is this place connected to other places, and what are the consequences?
- What changes are happening in this place, and what is causing them?
- What are your perceptions of other places, and on what are these perceptions based?
- What are other people's perceptions of your place, and are they accurate?
- How are places you know represented in popular music, films or TV shows? (see Kruse, 2004, for an example using the music of the Beatles).

Improving my place

- How could the liveability of this place be improved, and for whom?
- How can people's access to services and facilities be improved?
- How is development regulated and controlled, and by whom? Does the public have any power over development decisions?
- What organisations in this place are addressing local issues, and what motivates them?
- Are there any significant differences of opinion on what changes are needed in your place?

Bigger questions

- Where do you think you belong, and what does this mean to you?
- How does where you live influence your life?
- How does a place-based social, economic or environmental organisation in your area utilise 'place'?
- How could you use place analytically to answer a question?
- Can you make a generalisation based on a study of some aspect of places within your area?
- How do some places come to have a role in social, ethnic, religious or political conflict?
- Does place matter?

Useful reading

Cresswell, T. (2008). Place: Encountering geography as philosophy. *Geography*, *93*(3), 132–139. https://doi.org/10.1080/00167487.2008.12094234

Freeman, D., & Morgan, A. (2014). Teaching about places. *Teaching Geography*, *39*(3), 94–98. https://portal.geography.org.uk/journal/view/J078291

Freeman, D., & Morgan, A. (2017). Place and locational knowledge. In M. Jones (Ed.), *The handbook of secondary geography* (pp. 120–133). Geographical Association.
Larsen, T. B., & Harrington Jr, J. A. (2018). Developing a learning progression for place. *Journal of Geography, 117*(3), 100–118. https://doi.org/10.1080/00221341.2017.1337212
An American view of place.
Murphy, A. B. (2018). *Geography: Why it matters.* Polity Press. See chapter 3 on Place.
Oberle, A. (2020). Advancing students' abilities through the geo-inquiry process. *Journal of Geography, 119*(2), 43–54. https://doi.org/10.1080/00221341.2019.1698641
The use of inquiry to examine an issue in the local community.
Rawling, E. (2018). Reflections on progression in learning about place. *Journal of Geography, 117*(3), 128–132. https://doi.org/10.1080/00221341.2017.1398772
Table 2 in this article describes three different ways of studying places (descriptive, social constructionist and phenomenological), and the learning experiences and outcomes associated with each.
Rawlings Smith, E., Oakes, S., & Owens, A. (2016). *Changing places.* Geographical Association.
A guide to studying places written for A Level geography in England.
Taylor, L. (2005). Place: An exploration. *Teaching Geography, 30*(1), 14–17. https://portal.geography.org.uk/journal/view/J004378

References

Adams, P. C. (2017). Place. In D. Richardson (Ed.), *The international encyclopedia of geography.* John Wiley. https://onlinelibrary.wiley.com/doi/10.1002/9781118786352.wbieg0441
Agnew, J. (2005). Space: Place. In P. Cloke & R. Johnston (Eds.), *Spaces of geographical thought: Deconstructing human geography's binaries* (pp. 81–96). SAGE.
Agnew, J. A. (2011). Space and place. In J. A. Agnew & D. N. Livingstone (Eds.), *The SAGE handbook of geographical knowledge* (pp. 316–330). SAGE.
Amin, A. (2004). Regions unbound: Towards a new politics of place, *Geografiska Annaler: Series B, Human Geography, 86*, 33–44. https://doi.org/10.1111/j.0435-3684.2004.00152.x
Bambra, C. (2017). *Health divides: Where you live can kill you.* Policy Press.
Bambra, C. (2018). Placing health inequalities: Where you live can kill you. In V. A. Crooke, G. J. Andrews, & J. Pierce (Eds.), *Routledge handbook of health geography* (pp. 28–36). Taylor & Francis.
Beer, A., McKenzie, F., Blažek, J., Sotarauta, M., & Ayres, S. (2020). *Every place matters: Towards effective place-based policy.* Taylor & Francis.
Brooks, S. (2017). Coastal resilience and vulnerability: Storm impacts, extreme weather and regional variability in the UK, winter 2013–14. *Geography, 102*(2), 60–70. https://doi.org/10.1080/00167487.2017.12094011
Bustin, R. (2011a). Thirdspace: Exploring the 'lived space' of cultural 'others'. *Teaching Geography, 36*(2), 55–57. https://portal.geography.org.uk/journal/view/J002674

Bustin, R. (2011b). The living city: Thirdspace and the contemporary geography curriculum. *Geography*, *96*(2), 60–68. https://doi.org/10.1080/00167487.2011.12094312

Bustin, R. (2019). Investigating lived space: Ideas for fieldwork. *Teaching Geography*, *44*(1), 17–19. https://portal.geography.org.uk/journal/view/J004232

Casey, E. S. (2018). Foreword. In J. Malpas (Ed.), *Place and experience: A philosophical topography* (2nd ed., pp. viii–xiv). Routledge.

Creswell, T. (2014). Place. In R. Lee, N. Castree, R. Kitchin, V. Lawson, A. Paasi, C. Philo, S. Radcliffe, S. M. Roberts, & C. W. J. Withers (Eds.), *The SAGE handbook of human geography* (vol. 1, pp. 3–21). SAGE.

Cresswell, T. (2015). *Place: An introduction* (2nd ed.). Wiley Blackwell.

Government of Victoria, Department of Premier and Cabinet. (2020). *A framework for place-based approaches*. https://www.vic.gov.au/framework-place-based-approaches

Gregory, D., Johnston, R., Pratt, G. Watts, M. J., & Whatmore, S. (2009). *The dictionary of human geography* (5th ed.). Wiley-Blackwell.

Gregory, K. (2009). Place: The management of sustainable physical environments. In N. J. Clifford, S. L. Holloway, S. P. Rice, & G. Valentine (Eds.), *Key concepts in geography* (2nd ed., pp. 173–198). SAGE.

Hamnett, C., & Butler, T. (2011). 'Geography matters': The role distance plays in reproducing educational inequality in East London. *Transactions of the Institute of British Geographers*, *36*, 479–500. https://doi.org/10.1111/j.1475-5661.2011.00444.x

Hartshorne, R. (1960). *Perspective on the nature of geography*. John Murray.

Hill, R. (2019). Exploring *terroir*: A sense of place in food and farming. *Geography*, *104*(1), 42–48. https://doi.org/10.1080/00167487.2019.12094061

Hille, K. (1999). 'Gendered exclusions': Women's fear of violence and changing relations to space. *Geografiska Annaler: Series B, Human Geography*, *81*(2), 111–124. https://doi.org/10.1111/j.0435-3684.1999.00052.x

Holden, J. (2011). *Physical geography: The basics*. Routledge.

Horvath, R. J. (2004). The particularities of global places: Placemaking practices in Los Angeles and Sydney. *Urban Geography*, *25*, 92–119. https://doi.org/10.2747/0272-3638.25.2.92

Johnston, R., Manley, D., Pattie, C., & Jones, K. (2018). Geographies of Brexit and its aftermath: Voting in England at the 2016 referendum and the 2017 general election. *Space and Polity*, *22*(2), 162–187. https://doi.org/10.1080/13562576.2018.1486349

Jones, A. L. (2017). Regenerating urban waterfronts—Creating better futures—From commercial and leisure market places to cultural quarters and innovation districts. *Planning Practice & Research*, *32*(3), 333–344. https://doi.org/10.1080/02697459.2016.1222146

Jones, M. (2010). Limits to 'thinking space relationally'. *International Journal of Law in Context*, *6*(3), 243–255. https://doi.org/10.1017/S1744552310000145

Kramer, M. R., Black, N. C., Matthews, S. A., & James, S. A. (2017). The legacy of slavery and contemporary declines in heart disease mortality in the U.S. South. *SSM – Population Health*, *3*, 609–617. https://doi.org/10.1016/j.ssmph.2017.07.004

Kruse II, R. J. (2004). The geography of the Beatles: Approaching concepts of human geography. *Journal of Geography*, *103*(1), 2–7. https://doi.org/10.1080/00221340408978566

Kuyvenhoven, J., & Boterman, W. M. (2021). Neighbourhood and school effects on educational inequalities in the transition from primary to secondary education in Amsterdam. *Urban Studies*, *58*(13), 2660–2682. https://doi.org/10.1177/0042098020959011

Kwaymullina, A. (2005). Seeing the light: Aboriginal law, learning and sustainable living in country. *Indigenous Law Bulletin*, *6*(11), 12–15. http://www.austlii.edu.au/au/journals/IndigLawB/2005/27.html

Lai, K. P. Y. (2009). New spatial logics in global cities research: Networks, flows and new political spaces. *Geography Compass*, *3*(3), 997–1012. https://doi.org/10.1111/j.1749-8198.2009.00232.x

Little, S., & Derr, V. (2018). The influence of nature on a child's development: Connecting the outcomes of human attachment and place attachment. In A. Cutter-Mackenzie-Knowles, K. Malone, & E. B. Hacking (Eds.), *Research handbook on childhoodnature*. Springer. https://doi.org/10.1111/j.0435-3684.2004.00152.x

Malpas, J. (2006). The forms of water: In the land and in the soul. *Transforming Cultures eJournal*, *1*(2), 1–8. https://doi.org/10.5130/tfc.v1i2.257

Malpas, J. (2012). Is there an ethics of place? *Localities*, *2*, 7–31. https://citeseerx.ist.psu.edu/document?repid=rep1&type=pdf&doi=632129db580ce74aa59cec6eaa38263a25c31b56

Malpas, J. (2018). *Place and experience: A philosophical topography* (2nd ed.). Routledge.

Martin, R. (2001). Geography and public policy: The case of the missing agenda. *Progress in Human Geography*, *25*(2), 189–210. https://doi.org/10.1191/030913201678580476

Massey, D. (2002). Globalisation: What does it mean for geography? *Geography*, *87*(4), 293–296. https://www.jstor.org/stable/40573762

Pearcy, M. (2020). 'The Most Insidious Legacy'—Teaching about redlining and the impact of racial residential segregation. *The Geography Teacher*, *17*(2), 44–55. https://doi.org/10.1080/19338341.2020.1759118.

Rediscovering Geography Committee. (1997). *Rediscovering geography: New relevance for science and society*. National Academy Press.

Relph, E. (2017). Place and connection. In J. Malpas (Ed.), *The intelligence of place: Topographies and poetics* (pp. 177–204). Bloomsbury.

Rose, D. (2005). An indigenous philosophical ecology: Situating the human. *The Australian Journal of Anthropology*, *16*(3), 294–305. https://doi.org/10.1111/j.1835-9310.2005.tb00312.x

Roy, A., & Lane, S. (2003). Putting the geomorphology back into fluvial geomorphology: The case of river meanders and tributary junctions. In S. Trudgill & A. Roy (Eds.), *Contemporary meanings in physical geography: From what to why?* (pp. 103–125). Hodder Education.

Sayer, A. (2000). *Realism and social science*. SAGE.

Slater, T. (2013). Your life chances affect where you live: A critique of the 'cottage industry' of neighbourhood effects research. *International Journal of Urban and Regional Research*, *37*(2), 367–387. https://doi.org/10.1111/j.1468-2427.2013.01215.x

Smailes, P. (2000). The diverging geographies of social and business interaction patterns: A case study of rural South Australia. *Australian Geographical Studies*, *38*, 158–181. https://doi.org/10.1111/1467-8470.00109

Smith, C., Parr, N., & Muhidin, S. (2019). Mapping school's NAPLAN results: A spatial inequality of educational outcomes in Australia. *Geographical Research*, *57*(2), 133–150. https://doi.org/10.1111/1745-5871.12317

Smith, S. (2001). Residential segregation: Geographic aspects. *International Encyclopedia of the Social & Behavioural Sciences* (2nd ed.), *20*, 544–548.
Tammaru, T., Marcińczak, S., Aunap, R. van Ham, M., & Janssen, H. (2020). Relationship between income inequality and residential segregation of socioeconomic groups. *Regional Studies*, *54*(4), 450–461. https://doi.org/10.1080/00343404.2018.1540035
Walker, R. (2020). Wrestling with capital and class in US cities. *International Journal of Urban and Regional Research*, *44*(3), 546–548. https://doi.org/10.1111/1468-2427.12784
Wilke, H., & Budke, A. (2019). Comparison as a method for geography education. *Education Sciences*, *9*(3), 225. https://doi.org/10.3390/educsci9030225

Chapter 5

Thinking geographically

Space

Space is often seen as the core geographical concept, and at times geography has been defined as a spatial subject. It is also perhaps the hardest concept to explain clearly, as some of the writing about it in recent years has been complex and hard to follow. One reason for this is that in the literature on space, the term can have several different meanings. As Audrey Kobayashi writes:

> For students of the subject [space] is a bewildering term, sometimes depicted as an absolute product of nature with a predetermined structure, sometimes as a metaphor that is as changeable as the imagination, sometimes as an abstract concept that defies either specificity or logic. Often it is used uncritically by geographers—as 'space and place'—based on unquestioned assumptions that its use will be understood by readers. Such assumptions make understanding space even more difficult.
>
> (Kobayashi, 2017, p. 1)

This chapter explains some of the ways in which space has been conceptualised by geographers and then discusses a number of subsidiary spatial concepts, such as location, spatial distribution and spatial organisation, as it is these concepts that students can use and apply to understand and explain their world. The chapter concludes with some comments on spatial thinking and its applications.

Conceptualising space

How we think about space is important because it will affect how we use spatial thinking to try to understand our world. One approach is through the four ways of conceptualising space described below.

Absolute space

Absolute space is the material or physical space that is the surface of the Earth. It is the space we see from the air, extending in all directions and without

DOI: 10.4324/9781003376668-6

boundaries, but has little direct influence on human behaviour. This space is perceived as a container of objects created by nature or people whose positions can be fixed geometrically by map coordinates or latitude and longitude, or by their location relative to other objects, and whose spatial relationships with each other can be determined by measures of direction and distance. This is the space portrayed in topographical maps; maps of territorial units such as property boundaries, administrative areas and land use zones are also based on absolute space. Physical geography and Geographic Information Systems mostly adopt this concept of space.

Relative space

Relative space is based not on location and physical distance, but on factors such as the time and cost of moving between places or communicating with people in other places. Stuart Elden describes it in this way

> Other writers have focused on how space appears dependent on other things. On this view, space and our experience of it is relative to other things—to time, to cost or other economic factors, to social interaction, or the way in which we cognize or perceive it.
>
> (Elden, 2009, p. 264)

This form of space is produced by the construction of roads, railways, ports, airports and telecommunications and the invention of new forms of transportation and communication. It is constantly changing as a result of improvements in these technologies and investment in new infrastructure by governments and business. It is the space that influences individual and business behaviour, because it is mostly about cost and time, and it could be described as socially produced space.

Cognitive or perceived space

Elden, in the quotation above, mentions cognition and perception, but these are sometimes separated from relative space into the concept of *cognitive* or *perceived space*. This is the idea that space is subjective—that it is how we perceive and experience it—and varies between people, between societies and over time. For example, the perception of space by a globally oriented New York executive will be very different to that of the migrant worker from Latin America she employs in her apartment, or a person in an isolated village in the Highlands of Papua New Guinea, even though the latter may now own a mobile phone. Another example is how perceptions of distance differ between Australians, who are used to having to travel long distances between places, and Europeans (including the British) for whom distances between towns and cities are much smaller. As Rob Kitchen writes, 'We might occupy absolute space, but we live in cognitive space' (Kitchen, 2009, p. 268).

Cognitive space can be explored in schools by asking students to draw a mental map of their neighbourhood, the places they visit, their country or the world, as this will reveal their perceptions and use of space, and how these perceptions differ between them.

Relational space

Relational space is another idea about how to conceptualise space, but not an easy one to explain. It differs from relative space in that the significance of where something or someone is located relative to the location of other things and people depends on the relationships between them. For example, the strength of the connections between world cities is not just determined by the time and cost of transporting commodities, people and information between them but also by the density of the inter-city relationships between firms and people. These relational networks make these cities spatially much closer to each other than they are to small towns within their own country. As these relationships change, so do the spatial connections between cities, so that relational space is always changing, unlike absolute space which is constant. This influence of relationships is demonstrated in patterns of international trade, where physical distance has less effect on the volume of trade between countries than a common language, a former colonial relationship, a common currency or membership of a trading bloc.[1]

Conclusion

What can one conclude from this brief and selective review of ways of thinking about space? At the end of his own much longer and deeper review of the concept, Kitchen has this comment:

> It is fair to say that the conceptions of space outlined [in this article] ... are today all in use by geographers around the world. For example, absolute conceptions of space still predominate in spatial science and GISc and relative conceptions of space are popular with radical and feminist geographers. While new conceptual thinking is evolving all the time, as some geographers seek ever-more sophisticated ways to think about and analyze the world, older ideas persist rather than simply being replaced.
>
> (Kitchen, 2009, p. 274)

However, there are differences between physical and human geography in how space is conceptualised and used. In a book on key concepts in geography, Martin Kent writes that: 'There is no doubt that physical geographers have generally been less willing or less able than their human counterparts to adopt a predominantly spatial emphasis to their studies and research' (Kent, 2009, p. 98). He argues that physical geographers have tended to focus on processes and time rather than patterns and space, and where space has been considered, it is as absolute space.

These different ways of conceptualising space may seem somewhat theoretical and lacking practical application, but they inform how we think about location and spatial distributions, which are concepts that students can readily use, as explained in the following sections.

Locational concepts

Location is a fundamental element of space and is described by a range of concepts—absolute location, relative location, distance, time-space convergence, accessibility, centrality, proximity and remoteness. These are often absent from university texts on human geography, perhaps because they are seen as out-of-date or too mundane. Yet in real estate, retailing and public services such as schools and health, location is an important consideration. For example, the value of a house in a market economy depends partly on the perceived social status of where it is located. Furthermore, where things are located can have an influence on what they are like, and where people are located can have an influence on their lives, so these are still important ideas that students should understand.

Absolute and relative location

Absolute location is the unique location of a place as described by latitude and longitude, as in absolute space, while relative location is location in relation to other things, as in relative space. Note that place and location are not the same; while a place has a location, it also has distinctive characteristics or attributes, and complex meanings for people, so a place is much more than just a location. Relative location is generally more significant than absolute location. For example, the climate of a place is influenced not only by its latitude, which is part of its absolute location, but also by its location relative to land and water masses. Relative location affects businesses through their access to suppliers, markets, infrastructure, labour and information, but unequally between different types of business, as larger firms are more able to source components and services from distant places than smaller ones. Relative location affects individuals through their access to educational and employment opportunities, as mentioned in the discussion of place. Isolated locations distant from major centres are likely to provide fewer opportunities for both businesses (unless they are tied to the location of resources) and individuals, than locations in or close to large cities.

Accessibility

Accessibility, another locational concept, is closely related to relative location, because it is the ease with which people can travel to where employment, shopping, recreation or services such as health are located, and organisations can access the suppliers, services and information they need. It is important to recognise, however, that accessibility varies enormously between people in the same location. Accessibility for people with a physical disability, or who are dependent on public transport, or are caring for others, or feel vulnerable or unwelcome in public spaces, is likely to be constrained compared with others in the population. However, the need to travel for some services is being reduced by the use of telecommunications. For example, in writing this, I can explore the relevant literature from my home through a computer and the Internet and don't have to travel to a library, as I once had to. Furthermore, many of the books I buy are purchased online from suppliers in the UK, rather than from local bookshops. Changes such as these are having a growing and disruptive impact on retailing and consequently on retailing places. But many people in the world do not have access to online services, while the elderly who might benefit the most from them are often unable to learn them or afford them.

A very different example of the effects of accessibility comes from the programs of World Bicycle Relief (https://worldbicyclerelief.org), which operates in rural communities in South America, Africa, Sri Lanka and Southeast Asia. The program provides bicycles to people, particularly girls and women, so that they can better access education, healthcare and jobs. Instead of having to walk everywhere, children and adults can now ride a bicycle. This has increased school attendance, the number of patients that healthcare workers can visit and the quantity of produce a farmer can deliver to a market, and also improved the standing of women in the household. The work of the organisation illustrates very well the ways that distance does still constrain the lives and opportunities of many people throughout the world, and how a simple item of technology can make a significant difference.

Centrality

Centrality is the extent to which a location is in the centre of the spatial distribution of population, customers, businesses and employment. The central business district (CBD) of most cities, except for some in the USA, has high centrality because of its accessibility from the whole urban area. As a result, land values and building densities are very high (Figure 5.1). Similarly, rural towns are central to the areas they service, and those with higher centrality are tending to draw business away from places with lower centrality as increased mobility makes the former more accessible to rural customers. Large shopping malls are also located where they are central to and accessible by

Figure 5.1 Central New York. Has the 2020–2021 Pandemic reduced the attractiveness of CBDs and their dense concentrations of people?

Source: Author.

the maximum number of customers, and in some cases, this location may be outside a town.

However, the pull of the centrality of the CBD has been questioned in recent years. One reason is that in sprawling urban areas separate cores have developed away from the original CBD, often with easier accessibility because of the construction of new expressways. Another is that there has been a revival of the CBD as a residential area in many cities, typically for a mixture of counter-cultural people attracted by cheap former industrial buildings; higher-income people in creative industries such as the arts, design, media and technology, attracted by gentrified housing and the inner-city life style; and people employed in CBD firms. The COVID-19 pandemic has also had an effect on many cities. During lock-downs, where people were required to work from home if that was possible, the daily CBD workforce was drastically reduced, and many people discovered that they preferred to work from home for at least part of each week. If enough of these people continue to work from home, CBDs will lose part of their daytime population, reducing the money spent at local businesses and on office rentals. The pandemic also accelerated the growth of food and grocery delivery services in cities, which take food to where people live rather than consumers having to travel to restaurants, cafés and shops. Both these changes affect where things are located in cities, as well as the nature of work (Richardson, 2022).

Proximity

Proximity is about closeness to the services, people and knowledge important to a business, organisation or individual. For some types of business, especially financial and corporate legal firms, face-to-face access to both customers and specialist services is vital, and they tend to cluster together in the centre of major cities. Locational proximity is enhanced by firms' organisational or relational proximity, defined as 'the extent to which relations are shared in an organizational arrangement, either within or between organizations' (Boschma, 2005, p. 65).

Remoteness

Remoteness, the final locational concept, is about places which are relatively far from major population and economic centres. It has significant effects on people's economic opportunities and their access to education and other services. In health planning, the concept is used to identify areas which have the lowest accessibility to essential services.

Distance and time-space convergence

Distance is also important in our daily lives, because it constrains what we are able to do. We are likely to visit close places more frequently than distant ones, simply because of the time and cost in travelling longer distances. Distance influences our knowledge of other places, but this is also affected by our relationships with them and the information we gain from the media. We may know quite distant countries because of historical ties, or family connections, or because we compete with them in sport, and be less aware of closer countries with which we don't have these connections.

How distance is perceived and experienced is partly a function of the effort needed to overcome it, which depends on the infrastructure and technology that links places, and the way these are managed by businesses and governments. Improvements in transport and communication systems have greatly reduced the time and cost taken to move people, goods and information between places, in a process called time-space convergence or time-space compression. However, these improvements have been greater in some places than in others. For example, in physically large countries like the USA and Australia, the major cities are closer to each other in travel time by air than they are to nearer towns only accessible by land transport. Similarly, at the global scale, many regions in lower-income countries have had little benefit from improved communications.

The effects of time-space convergence or compression have also been highly unequal between people. Massey (1991) comments on the 'power geometries' involved in convergence. She argues that there are differences between people in their degree of movement and communication and in their degree of control and initiation. Some are both doing the moving and communicating and are in a position to control and benefit from it. There are others who are doing the moving but are not in control and who gain fewer benefits; this group includes illegal migrants, imported low-cost labour and refugees. And there are the large numbers unable to move or prevented from moving. Most of the world's population cannot afford to fly or own a car, and about half do not have access to the Internet. Time-space convergence is both geographically and socially very uneven.

Does distance still matter?

The convergence or compression of time and space has led some to contend that geography is no longer relevant, an argument known as the 'death of geography' thesis. This claims that as knowledge is now instantly accessible anywhere in the world through information and communications technologies, people no longer need to meet face-to-face to obtain it. All locations are consequently equal in their access to knowledge, proximity is no longer a locational determinant, and a non-central location is no longer a disadvantage. This claim ignores several things. One is the existence of a digital divide, with some places, and some people, not having good access to the Internet and therefore to knowledge (Roser et al., 2015; Warf, 2019). Another is the difference between codified and tacit knowledge. The former is knowledge that is recorded and consequently can be communicated widely in printed or electronic forms. The latter is knowledge that is in people's heads, and this is most effectively communicated face-to-face. Tacit knowledge is important in activities involving learning and innovation, and its effective transmission also involves personal relationships and trust. As Kevin Morgan writes, 'something gets lost, or degraded, when individuals and organizations communicate at a distance' (Morgan, 2004, p. 8). This is compounded when the distant places have different cultures, managerial attitudes and workplace practices.

Another way of demonstrating that geography is not dead is in studies of the influence of distance on people's online social networks. It might be expected that online connections would be unaffected by distance because their cost and time does not depend on the relative location of users. However, an analysis of data on ten million active users of a Spanish social network concluded that:

- There is a marked preference to connect to spatially closer users, so distance influences with whom users interact.
- Teenagers under 19 appear to be more constrained in their social networks than older people, presumably because they are less mobile geographically (Laniado et al., 2018).

Similarly, a study of Facebook social networks in the New York metropolitan area concluded that:

> Even in an era of increasing reliance on communications technology, we find that physical distance remains an important determinant of social connections. Yet, we also provide evidence of the important role of transportation infrastructure in forming and maintaining urban social connections by showing that social networks are distributed along public transportation routes and that social connectedness between locations declines more in travel time than it does in physical distance.
>
> (Bailey et al., 2020, n.p.)

What these studies show is that people's online social connections are similar to and complement their physical social networks and are influenced by distance.

Agglomeration

A major problem with the 'death of geography' thesis is that it is unable to explain the continuing growth of cities where activities cluster or agglomerate together because of the advantages of proximity. Allen Scott and Michael Storper (2015, p. 6) describe agglomeration as 'the basic glue that holds the city together'. For firms the advantages of clustering together include:

- Access to a large pool of labour, and one that is increasingly subdivided into specialisms. This makes it easier for firms to obtain the specific skills they require.
- Access to specialised equipment and the technical services to maintain it.
- Availability of suppliers of intermediate goods and buyers of finished products.
- Shared infrastructure, such as transport, energy and buildings.
- Bringing 'people and firms together in close proximity is thought to create fertile social conditions for experimentation, learning and innovation. ... Geographic concentration facilitates the ready circulation and exploitation of both codified and "tacit" knowledge' (Barnes and Christophers, 2018, p. 244).

Scott and Storper describe agglomeration as:

> a mechanism of sharing, matching and learning. Sharing refers to dense local interlinkages within production systems as well as to indivisibilities that make it necessary to supply some kinds of urban services as public goods. Matching refers to the process of pairing people and jobs, a process that is greatly facilitated where large local pools of firms and workers exist. Learning refers to the dense formal and informal information flows (which tend

> to stimulate innovation) that are made possible by agglomeration and that in turn reinforce agglomeration.
>
> (Scott and Storper, 2015, p. 6)

These advantages suggest the following generalisation:

> Because of the advantages of agglomeration, economic activities tend to cluster in space, at all scales from the local to the global, unless tied to the location of natural resources or dispersed customers.

This is a generalisation that students could apply to make sense of new contexts or to forecast future spatial trends. They might also discover economic activities that do not follow the generalisation and try to work out why and whether it then needs to be amended.

We may conclude that although technology has enabled many economic functions to be relocated both within and between countries, and a range of services to be outsourced to lower-cost places, distance continues to influence where things are located and who we communicate with. Geography is not dead.

Conclusion

To illustrate some of these locational ideas, Figure 5.2 is a map of the cost of one-bedroom flats in London. The lightest shading represents areas for which there is insufficient data and the darkest those with the highest rents. High rent areas are clustered around the centre of London, demonstrating the effects of relative location, centrality, proximity and accessibility. However, after falling with increasing distance from the centre, rents rise again, showing that relative location is not the only influence.

This discussion of location concludes with Tobler's First Law of Geography (TFL), which states that 'Everything is related to everything else, but near things are more related than distant things'. This was published in 1970 and is rejected by many human geographers as a relic of positivist spatial analysis. However, it has refused to die. It was reviewed by a panel of geographers in 2003, and while several argued that it did not qualify as a law, 'the consensus was reached that TFL is applicable to the real world' (Foresman and Luscombe, 2017, p. 981). A review of the Law by Hans Westlund concluded that:

> The transformation from a manufacturing-industrial economy to a global knowledge economy has meant that the first law of geography has lost its explanation power for certain activities. The importance of proximity has decreased in a number of areas at the same time as mobility has increased [as a consequence of the] increased mobility of production factors and decreased costs of transportation. … There is however, at least one area where 'near things are more related than distant things,' and that is the dense

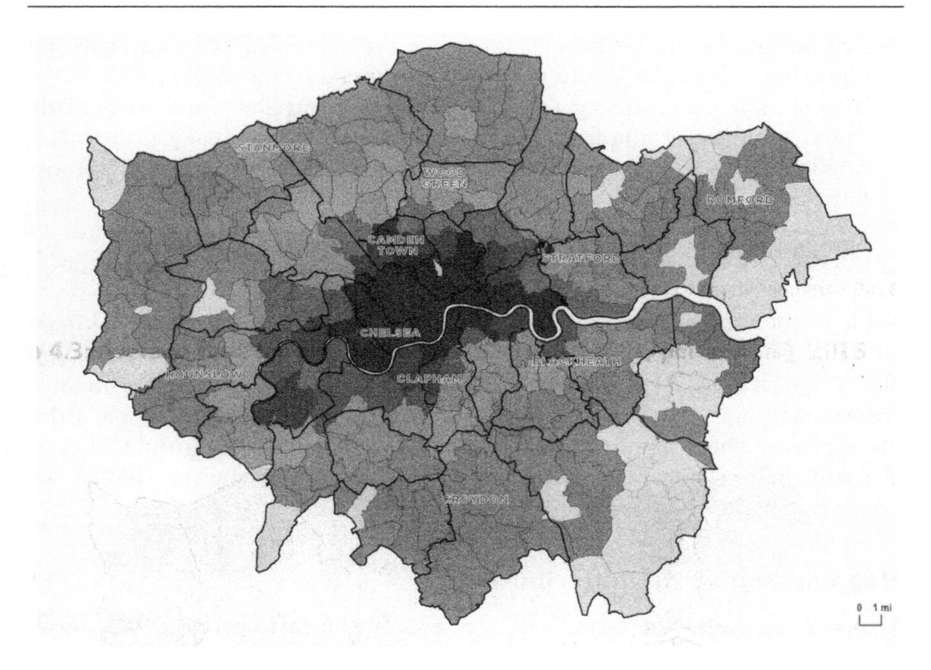

Figure 5.2 Average rent for one-bedroom flats in London, 2015. © London and Partners.

Source: GLA Economics, 2016, p. 149.

creative environments where new specialized, tacit knowledge is born and where existing tacit knowledge is transformed and commercialized.

(Westlund, 2013, p. 923)

Diffusion

Diffusion is a complex spatial concept because it combines space and time and is about the spread of something from its origin across space and over time. It could be the spread of a new fashion, a new technology, successive outlets of an expanding retail company, a plant, animal or human disease, or a financial collapse. The aim of diffusion research is to identify the pattern of spread, explain it and use that understanding to forecast its future distribution and inform measures to control it. An example is the early diffusion of COVID-19 in Germany from mid-February to mid-March in 2020 (Kuebart and Stabler, 2020). The researchers found that, instead of spreading outward from an original source of the infection to the rest of the country, as anticipated, the disease appeared very quickly all over Germany. They found three spatial patterns:

- Regional outbreaks from super-spreading events, such a major carnival or a nightclub, where the spread was mostly confined to the local region.
- Infections from returning tourists or business travellers, and particularly from tourists returning from Italy and Austria. The latter caught the infection in super-spreading places such as ski resorts and then spread it to most districts of Germany.
- Local outbreaks in closed communities, such as when someone brought an infection into an aged care facility.

None of these patterns follow the usual one of spatially outward spread from an initial point. Instead, the researchers point to the role of super-spreading places in which people were in close proximity, such as ski resorts, nightclubs, festivals and aged care homes, and the movement patterns of people from those places. These patterns reflect the mobility behaviour of groups of people, and their relationships with other people, which is why the title of the research includes the term 'socio-spatial'.[2]

Mapping and spatial distributions

Mapping is a method of portraying space and it is central to geography because:

> Maps have the power to turn the abstract ideas, which we form in our heads, into visual reality. Only a handful of people will ever actually see the UK complete from space. Even fewer can expect to see the whole world in one go except as a map. Maps and plans have a potential for radically extending our understanding by portraying the layout and organisation of the school, revealing the network of roads in a town or region or showing the distribution of natural vegetation such as forests and grasslands. Maps contextualise information within defined spatial boundaries, allowing us to make comparisons, formulate plans and develop generalisations. The identification and analysis of patterns, processes and relationships stands at the heart of geography.
>
> (Bridge, 2010, p. 116)

A spatial distribution is the pattern produced by mapping information about a specific phenomenon. Visualising data as a spatial distribution can be a powerful way of making sense of the phenomenon being mapped, but to extract the full value from a spatial distribution, students should progress from describing it to use it to raise questions, evaluate its significance and implications, explain it and use it analytically to answer questions.

Using a spatial distribution to analyse

Chapter 2 described mapping as a type of analysis that could reveal patterns in data that are not immediately apparent in a table and which may suggest ways of explaining the data. Figure 5.3 provides an example. It maps the total

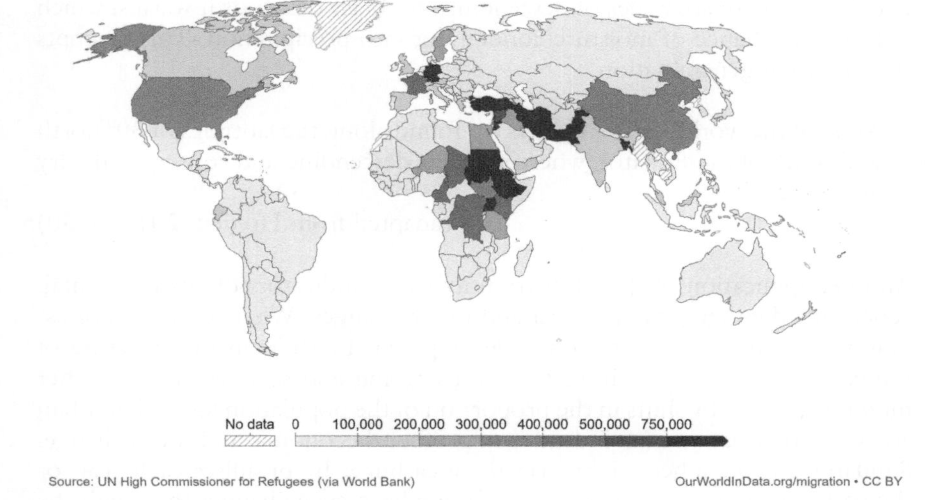

Figure 5.3 Number of refugees by country or territory of asylum, 2020.

Source: Ortiz-Ospina et al. (2022). https://ourworldindata.org/grapher/refugee-population-by-country-or-territory-of-asylum. Reproduced under Creative Commons.

number of refugees in 2020 by the country in which they have sought refuge and reveals clusters of refugees in central Africa, Bangladesh and West Asia. Outside these regions only Germany had large numbers of refugees, despite popular perceptions of Europe as a major refugee destination. This pattern might prompt students to ask questions such as:

- What explains these clusters?
- These countries are not wealthy, so how do they cope?
- Why does Germany have a large number of refugees, and not other Western countries?
- Why are there relatively few refugees in countries such as Australia and the USA, which take large numbers of international migrants each year?

Another map from the same source will tell students that the main countries that produced these refugees were Sudan, South Sudan, the Democratic Republic of Congo, Somalia, Myanmar, Afghanistan and Syria. From this, they might conclude that most refugees are living in countries next to their country of origin, so that refugee migration is mostly short-distance.

A map of world climates provides another example. Students could observe that the world's subtropical deserts are located on the western sides of the continents at around 30° north and south of the Equator. This repeated locational pattern suggests a common cause, and with some further investigation, students should discover that, because of the global patterns of atmospheric circulation, these are zones of descending air. As air descends it warms, which reduces the chance of moisture condensing into precipitation. This prompts the following generalisation

> Most of the world's major deserts are found along the latitudes of 30^0 north and south of the Equator, where the air is descending and consequently dry and clear
>
> (adapted from Holden, 2011, p. 30)

Another application of spatial distributions is as indicators of environmental, economic, demographic, political and other changes. Whereas an economist might measure national economic development by shifts in the structure of the economy from agriculture to manufacturing and services, a geographer might measure it by shifts in the proportion of the population located in urban areas, as urbanisation is an indicator of profound economic and social change. Similarly, a geographer might map the area burnt by bushfires each year, or the paths of tropical cyclones, to see if there have been changes that might be the result of climate change.

Evaluating the significance and implications of a spatial distribution

Spatial distributions are not just patterns on a map but have significant environmental, economic, social and political implications. An example of a spatial distribution that illustrates this is shown in Figure 5.4. This maps the pattern of human development by counties in the USA, as measured by an index that combines data on life expectancy, income per capita, school enrolment, percentage of high school graduates and percentage of college graduates. There is a clear spatial pattern in the distribution, with low scores in the south (except for Florida, which has a high level of retirement migration from other states) and higher scores in the north and west. This pattern could help to explain the geography of political affiliations and Presidential voting and a range of social and religious attitudes across the USA.

Explaining spatial distributions

Spatial distributions can be used to identify possible causal relationships through the method of map comparison. An example is to look for a relationship between vegetation and climate at a global scale by comparing maps of climatic types and major biomes. Similarly, students could compare a map of US Presidential votes with ones of per capita income, level of educational attainment and ethnicity to see if there is a relationship between them. However, it is important to understand that this method, like statistical analysis, can only suggest that there may be a cause-and-effect relationship; it does not prove

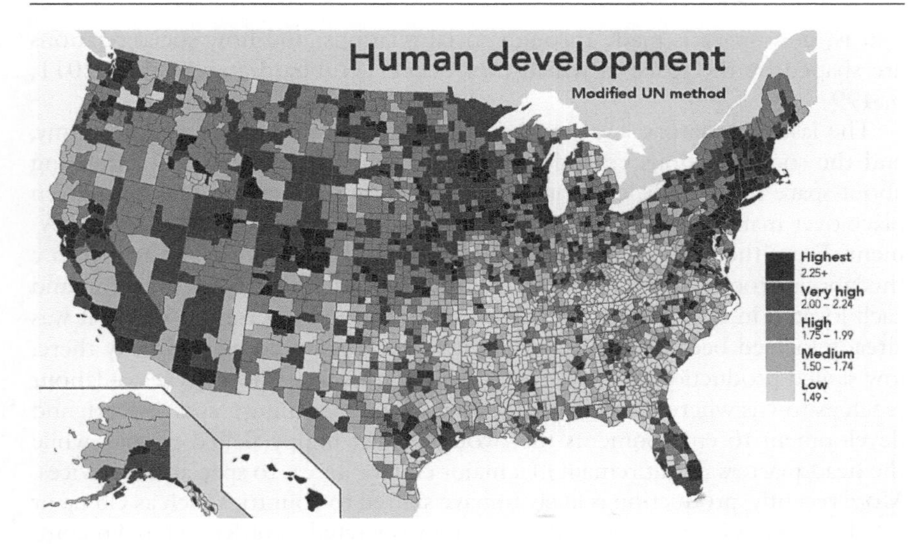

Figure 5.4 Human development in the USA by county, 2013.

Source: https://i.pinimg.com/originals/8c/1a/9d/8c1a9d3f1d04460ffee56bb071b80c94.gif. Reproduced with the permission of the creator, Sasha Trubetskoy.

one. To do that one has to identify the causal processes or mechanisms that produced the relationship.

Explaining a spatial pattern in human geography can be attempted at several levels, as explained in Chapter 2. For example, to explain the spatial distribution of different socioeconomic groups within a large city, researchers might look at people's varying preferences on where to live, which might be near their place of employment, or near relatives or people of the same ethnicity, or in suburbs with particular types of amenity. At a deeper level researchers might investigate how in a capitalist economy the property market constrains the choices of lower-income people by pricing them out of particular areas or discriminates against some ethnic groups, as mentioned in Chapter 4. They might go even deeper and argue that the spatial distribution of income groups is a product of social inequality, which in turn might be explained by capitalism or by the inadequate regulation of capitalism by governments. These explanations start with ones that focus on people's preferences and choices and then go progressively deeper into the structures that constrain those choices. Students might also examine whether the spatial distribution of socioeconomic groups, and particularly their degree of spatial segregation, helps to perpetuate their social inequality, as suggested in Chapter 4 in relation to educational outcomes. While the structure of society helps to explain a spatial pattern, does that pattern in turn help to explain the structure? Geographers have created the term 'spatiality' to describe this interrelationship, defining it as: 'A term that refers to how space and social relations are made through each other;

that is, how space is made through social relations, and how social relations are shaped by the space in which they occur' (Hubbard and Kitchin, 2011, p. 499).

The last explanatory level looked for answers in the society and economy, and the social relations embedded in them. This is another way of thinking about space relationally, and can be illustrated by the changes that have taken place over many years in the spatial distribution of types of industrial employment. From the 1960s, large firms in the UK and the USA began to separate the labour process into separate stages, each with different types of labour and each located in different places. Skilled production might remain where it was already located because the experienced workers required were already there, low skilled production work might be moved to places with reserves of labour (such as towns where the existing industries were declining), and research and development to environments that would attract highly skilled people, while the headquarters might remain in a major city for access to specialised services. More recently, production is likely to have shifted to countries such as China or Thailand. So as firms changed the ways they operated in order to remain competitive and profitable, they created new spatial patterns of employment. Where production was located in space (the spatial form of production) changed from a single place to places scattered across one nation, and then to locations scattered across the world, because the labour process of production changed.

Marxist geographers argue that this spatial pattern of continuing uneven development is an inevitable consequence of capitalism and necessary for its survival. This is because one way capitalist firms can maintain their competitiveness and profitability is to change the location of some of their activities, or outsource the supply of some products and services, to places with cheaper labour. In this process, some places gain and others lose employment and income, in a process of continuing change and disruption. Uneven development is therefore an inevitable consequence of capitalism. Others might say that uneven development is an inevitable consequence of technological change, as will be discussed in Chapter 7, but then isn't technological change driven by capitalism?

The organisation of space

Within each place, such as a home, a suburb, a city or a nation, is a space that is organised and planned in various ways. The space within your home is organised into areas for sleeping, washing, cooking, eating and leisure. The space within cities is zoned for specific activities, such as manufacturing, parks, shops and housing, and within it road and rail routes are planned and services such as hospitals and schools located. Other examples of the organisation of space are the designation of local government areas, the spatial pattern of sports leagues and associations, the establishment of administrative areas by government agencies and businesses and the areal divisions of organisations like Rotary.

The drawing of electoral boundaries is a particularly contentious type of spatial organisation in some countries, especially the USA, because they can be designed to favour one party or the other in a process called 'gerrymandering'. Gerrymandering is named after Elbridge Gerry, who as Governor of Massachusetts in 1812 redrew the state senate electoral districts to favour his party. One form of gerrymandering is the design of electoral districts by the party in power so as to group the maximum number of supporters of the opposite party into the smallest number of districts, leaving a larger number of districts in which the party in power has a chance of winning. Gerrymandered districts have very odd shapes and may combine areas that are not contiguous; Figure 5.5 shows historical examples from the USA. To prevent gerrymandering by politicians, many countries have established independent bodies to draw electoral boundaries based on specified criteria and to adjust them after each census in response to changes in the distribution of population.

In *The Production of Space*, Henri Lefebvre (1991) argued that different economic, social and cultural systems have or had different ways of organising space. He illustrated his ideas particularly with accounts of the form of cities, arguing, for example, that feudal cities were focussed on the palace of the ruler, while capitalist cities changed their form as capitalists moved production to places which were the most profitable, which in turn required ever-growing networks of transportation and communication. This is a Marxist interpretation, in which spatial form is produced by the mode of production and its social relations.

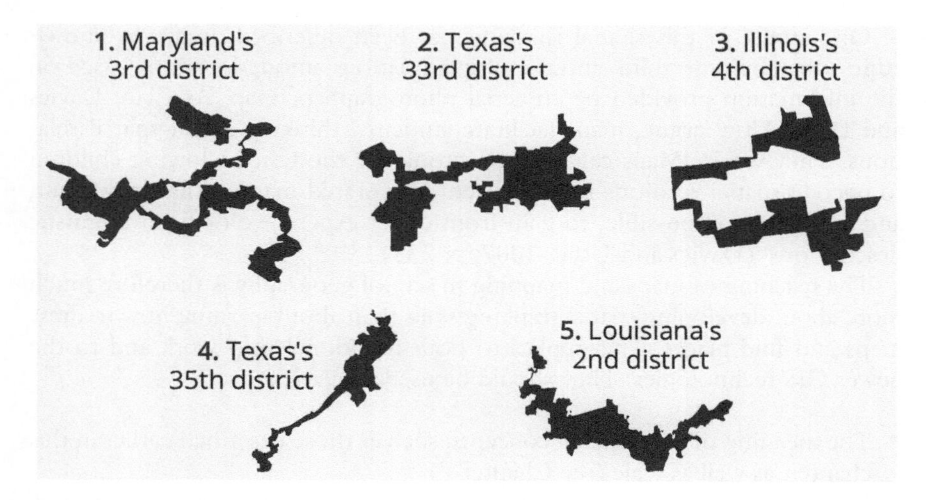

Figure 5.5 Gerrymandered Congressional Districts in the USA. The shape of the map for Maryland's third district has been nicknamed 'The Preying Mantis'.

Source: Statista, http://cdn.statcdn.com/Infographic/images/normal/21313.jpeg. Reproduced under Creative Commons.

Spatial thinking and its applications

As defined by Sarah Bednarz, spatial thinking:

> emphasizes language (knowing and using spatial concepts such as location, distance, scale); being able to understand spatial representations such as maps, graphics, and diagrams; and the application of these to problem solving, both personal and academic. This is related to the development of a spatial habit of mind. This is the predilection to think spatially and to apply the skills required to engage in reasoning with concepts of space and visual representations.[3]
>
> (Bednarz, 2018, p. 3)

The ability to think spatially is educationally important because it helps to develop children's spatial intelligence, a separate type of intelligence additional to mathematical and verbal (Ness et al., 2017). Spatial intelligence is important in everyday life but is also used in mathematics, several fields of science, architecture, engineering, urban planning and geography. Skill in spatial thinking is positively correlated with competence in mathematics and some branches of science, and in a recent article, Nicholas Judd and Torcel Klingberg (2021) report strong evidence that spatial cognitive training improves mathematical learning in children. Geography has a significant role to play in this training, and Lynn Liben (2017, p. 221) argues that 'geography education in general, and map education in particular, can have an important place in developing spatial thinkers'.

One aspect of this spatial thinking has been described as survey knowledge—the ability to think about multiple relations among locations based on the information provided by an aerial photograph or map. As Clare Davies and David Uttal argue, maps facilitate students' thinking about spatial relations. They write: 'Maps can become "tools for thought", allowing children to encode spatial relations in an efficient, integrated manner that is difficult, and sometimes impossible, to gain from direct experience or from linguistic descriptions' (Davies and Uttal, 2007, p. 233).

The teaching of maps and mapping in school geography is therefore much more about developing spatial thinking skills than about learning how to draw maps and find places. This applies to both traditional map work and to the newer GIS technologies. They should be used to teach:

- The meaning of basic spatial concepts, such as those described earlier in this chapter, as well as scale (see Chapter 7).
- How to interpret more complex spatial concepts, such as spatial patterns, clusters, associations and diffusion.

- How to evaluate the quality of spatial data.
- How to use maps and GIS to analyse, compare, explain, generalise and evaluate—in other words, to reason spatially.

In the context of this book, the last point is important, because it is about thinking geographically. Spatial visualisations can be used simply to portray information, but for geographical thinking, they should also be used to reason. That this may not be happening in schools is suggested by a study by Injeong Jo and Sarah Bednarz (2009) of the questions asked in four American high school geography textbooks. They found that of the minority of questions that were spatial, most asked about simple spatial concepts such as location, and there were much fewer about the more complex concepts or about reasoning. It is consequently important that teachers make up for the deficiencies in textbooks by using mapping and GIS in more challenging ways. Mary Fargher (2019), and Mary Fargher and Grace Healy (2020), demonstrate how GIS mapping can be used to teach the powerful geographical knowledge described in Chapter 2.[4] Using their methods, a GIS could be used to investigate issues such as:

- Trends in land clearance in a region, and their possible causes (see Chapter 10).
- Mapping areas of future flood risk under different assumptions about future rainfall events.
- Modelling the effects of climate change on the habitats of endangered animals.
- Measuring the spatial association between disadvantaged social groups and air pollution or other environmental hazards.

Other applications of spatial thinking, with or without the use of GIS technology, include:

- Finding the causes of diseases by mapping their incidence.
- Mapping areas at risk from sea-level rise and storm surges.
- Assisting the search for serial criminals by spatially profiling the most likely location of their next criminal actions.
- Locating schools and health services by mapping to identify areas that are currently under-serviced or are forecast to be areas of future demand.
- Deciding on the optimum location for a business activity. The need for this service sustains a number of national and international consultancy firms. Box 5.1 describes a simple example of the use of spatial thinking to decide on a location for a new activity.

Box 5.1 Locating a training ground for a football club

Richmond is an Australian Rules football club formed in 1885, based in the old inner suburbs just southeast of the Melbourne CBD. In 2008, the club began occasional out-of-season training at a location some 30 km north of its base, in an area of rapid population growth and young families, with the aim of attracting new members. However, the location was not successful, and the club approached a demographic and location analysis firm (.id) for advice. The firm reasoned that followers of a football club are generational, so an appropriate location for such an event would be an area into which the children and grandchildren of Richmond supporters were moving. Migration data revealed that these descendants were moving to the east and southeast of the metropolitan area and not to the north. So the firm recommended a location some 60 km to the southeast, in another area of rapid population growth. In collaboration with the local government of the region, the club hosts at least one open training session a year at the location, together with a football clinic for children, player signing sessions and other events. This has been much more successful in attracting new members.
This case study illustrates the concept of relational space. The second location was much further from the club's base than the first one, but it was connected to the base by much stronger relationships, through migration, than the first one.

Source: https://home.id.com.au/case-studies/richmond-football-club

Space and public policy

The concept of space underlies a variety of policies tried by governments at different levels to achieve particular aims. Below are some examples:

- In urban areas, development is often managed by creating zones in which only specified types of land use, economic activity and buildings are permitted. This is intended to prevent the juxtaposition of incompatible activities, such as noxious industries and housing, but is also used to protect the house values of nervous owners. Students could investigate whether zoning differs between the richer and poorer areas of a city and provides less protection for disadvantaged people from environmental and other hazards.
- Strategies to protect biodiversity may designate specific land and marine spaces in which harvesting is restricted and actions which damage the productivity of the environment are banned.

- To attract business investment to disadvantaged areas in a country, governments may provide tax concessions, cheaper credit, assistance in obtaining labour and other incentives for firms that locate in designated places.
- To provide alternative economic locations to the major centres in a country, governments may designate new growth centres and build the infrastructure required to make them attractive. In capitalist economies, these have not been very effective. Brasilia, established in 1960 as the new inland capital of Brazil, is an example. The city was meant to create a new growth node to counter the dominance of Brazil's two major cities, São Paulo and Rio de Janeiro, which are located on the coast. A detailed study of the spatial effects of Brasilia concluded that:

> Faced with agglomeration and other forces, even a massive development initiative of the scale [of Brasilia] is estimated to have had limited population impacts outside of Brasilia itself and approximately zero per capita income impacts on the spatial structure of Brazil's urban economy.
>
> (Grimes et al., 2017, p. 796)

Space and urban planning

Urban planning has been described as an application of spatial geographical thinking and analysis to practical problems, and many geography graduates have gone on to study planning and become planning professionals. Geography researches and teaches many topics that are relevant to planning, and can provide the understanding of the geographical complexity of urban spatial structure, and how this is changing, that can inform urban planning policies and practice. However, this contribution may be declining as human geography has become more critical, more theoretical, more cultural and more qualitative, and perhaps less relevant to planning issues.

A spatial perspective of the world

The British geographer Doreen Massey has written extensively about the difference between an historical or time perspective and a spatial perspective of the world. She argues that in an historical perspective, the countries and places of the world are viewed as being on the same trajectory of change and development as the 'developed countries' and differ only by their position on that trajectory. In a spatial view of the world, countries and places exist simultaneously in time, have distinctive characteristics, and may be following different trajectories. Space makes possible the co-existence of multiplicity and difference (Massey, 2005). Andrew Sayer (2000) uses a similar argument to highlight the significance of areal differentiation. He points out that national societies are not unitary objects, but differ internally across space, such that the experience of ethnicity, for example, will differ across a nation. Both Massey and Sayer draw attention to how spatial thinking emphasises the variety of and differences between places, regions and countries, without any suggestion that some are more advanced or more typical than others. They are just different.

What is the difference between space and place?

This chapter began with a comment on how the distinction between space and place is sometimes unclear, so it may help to try to differentiate between the two key terms. As a geographical concept, space is about location, distance, spatial distribution and spatial organisation, and their influence on the environment, people and societies. Place, on the other hand, is about the characteristics of the areas of the Earth's surface we identify as places and their influence on environmental and human processes and phenomena. Very simply, space is about 'where', and place is about 'what is there'. Problems arise, however, when spaces are described in the way that Chapter 4 defined places. So be aware that academic geographers often talk of spaces when we might think that they are really places. They also sometimes talk of spaces as generic types of places, such as 'spaces of fear', 'economic spaces', 'tourist spaces' and 'manufacturing spaces'.

Conclusion

This chapter has unpacked the rich concept of space into a number of smaller ideas and described how they can be used to portray, understand, explain and analyse the world. It discussed how to conceptualise and think about space, and how to use the different conceptualisations. The concept of space teaches us to think spatially, by recognising the influence of location and distance on people and places, by using spatial distributions to identify causes and consequences, by understanding how people organise and manage space and by recognising the areal differentiation of the world. But above all, it may teach us how the ways our society is structured influences our geography and how this geography may in turn influence our society.

Summary

- Space can be conceptualised as absolute, relative, perceived or relational.
- Location is a fundamental element of space and can be unpacked into more specific ideas such as absolute location, relative location, accessibility, centrality, proximity, remoteness, distance and time-space convergence.
- Despite major technological changes in transportation and communication technologies, locations are not equal in their accessibility,

especially to information and knowledge, or in their ability to attract and foster particular types of economic activity.

- Cities form and grow because of the economic advantages of agglomeration.
- Diffusion is a complex spatial concept that combines space and time and is about the spread of something from its origin across space and over time.
- Mapping is central to geography because it is a way that geographers portray spatial distributions.
- Spatial distributions are important because they have implications and consequences, need explanation and can be used to answer questions.
- The ways that space is organised for particular purposes can have significant economic, social and political consequences.
- The concept of space is central to spatial thinking and to the development of spatial intelligence.
- Spatial thinking, increasingly supported by GIS technology, can be used to answer questions and solve problems.
- The concept of space underlies a variety of public policies.
- Urban planning has been described as an application of spatial geographical thinking.
- Space makes possible the co-existence of multiplicity and difference.

How could you use this chapter in teaching?

Questions for students based on the content of this chapter include:

- How is my daily life influenced by relative location?
- How does the relative location of where I live influence what it is like?
- Are there any economic activities in my area that benefit from proximity?
- Do the friends I have on social media follow TFL?
- Is my knowledge of other countries influenced by their distance from me, my relationships with them, the media or some other factors?
- How many of the apps on my mobile phone use locational information, and for what purpose?
- Is the CBD nearest to me growing or declining, and if so, why?
- Has the Pandemic of 2020–2022 changed the daily lives of people I know?
- How are the sports and other activities in which I participate spatially organised?

- What is the spatial distribution of an environmental or social phenomenon in my area, and how can the pattern be explained?
- How are the electoral boundaries in my country or state determined, and how fair are the outcomes?
- How is the space of my suburb or area organised, and for what purposes?
- How is the Internet changing retailing and service provision in my place?
- Is distance a constraint?

Notes

1 For criticisms of relational space, see (Jones, 2010; Malpas, 2012).
2 For ideas on using COVID-19 to teach geographical thinking, see (Dolan and Usher, 2022; Sittner, 2021). For ideas on teaching a geography of health, see (Davies-Crane, 2021).
3 See also (National Research Council, 2006).
4 For other examples, see (Healy and Walshe, 2019; Hickman, 2021; Law, 2022; Larsen and Solem, 2022).

Useful readings

Bednarz, S. W. (2018). Spatial thinking: A powerful tool for educators to empower youth, improve society, and change the world. *Boletim Paulista de Geografia*, *99*, 1–20. https://publicacoes.agb.org.br/boletim-paulista/article/view/1458
Elden, S. (2009). Space I. *International Encyclopaedia of Human Geography*, 262–267. https://www.sciencedirect.com/science/article/pii/B9780080449104003205
Goldin, I., & Muggah, R. (2020). *Terra incognita: 100 maps to survive the next 100 years.* Century.
Contains maps and extensive commentaries on many geographical topics.
Kitchen, R. (2009). Space II. *International Encyclopaedia of Human Geography*, 268–275. https://www.sciencedirect.com/science/article/pii/B9780080449104011263
Morgan, J. (2003). Teaching social geographies: Representing society and space. *Geography*, *88*, 124–134. https://www.jstor.org/stable/40573831
Murphy, A. B. (2018). *Geography: Why it matters.* Polity Press.

References

Bailey, M., Farrell, P., Kuchler, T., & Stroebel, J. (2020). Social connectedness in urban areas. *Journal of Urban Economics*, *118*, 103264. https://doi.org/10.1016/j.jue.2020.103264
Barnes, T. J., & Christophers, B. (2018). *Economic geography: A critical introduction.* John Wiley.
Boschma, R. (2005). Proximity and innovation: A critical assessment. *Regional Studies*, *39*(1), 61–74. https://doi.org/10.1080/0034340052000320887

Bridge, C. (2010). Mapwork skills. In S. Scoffham (Ed.), *Primary geography handbook* (pp. 104–119). Geographical Association.

Davies, C., & Uttal, D. H. (2007). Map use and the development of spatial cognition. In J. M. Plumert & J. P. Spencer (Eds.), *The emerging spatial mind* (pp. 219–247). Oxford University Press.

Davies-Crane, G. (2021). Geography of health for key stage 3. *Teaching Geography*, *46*(3), 94–97. https://portal.geography.org.uk/journal/view/J004305

Dolan, A., & Usher, J. (2022). The geography of COVID-10: A case study in geoliteracy. *Teaching Geography*, *47*(1), 50–52. https://portal.geography.org.uk/journal/view/J004696

Fargher, M. (2018). WebGIS for geography education: Towards a geocapabilities approach. *International Journal of Geo-Information*, *7*(3), 111. https://doi.org/10.3390/ijgi7030111

Fargher, M., & Healy, G. (2020). Empowering geography teachers and students with geographical knowledge: Epistemic access through GIS. In N. Walshe & G. Healy (Eds.), *Geography education in the digital world: Linking theory and practice* (pp. 102–116). Routledge.

Foresman, T., & Luscombe, R. (2017). The second law of geography for a spatially enabled economy. *International Journal of Digital Earth*, *10*(10), 979–995. https://doi.org/10.1080/17538947.2016.1275830

GLA Economics. (2016). *Economic evidence base for London 2016*. Greater London Authority. https://www.london.gov.uk/sites/default/files/economic_evidence_base_2016.compressed.pdf

Grimes, A., Matlaba, V. J., & Poot, J. (2017). Spatial impacts of the creation of Brasilia: A natural experiment. *Environment and Planning A*, *49*(4), 784–800. https://doi.org/10.1177/0308518X16684519

Healy, G., & Walshe, P. (2019). Real-world geographers and GIS: Relevance, inspiration and developing geographical knowledge. *Teaching Geography*, *44*(2), 52–55. https://portal.geography.org.uk/journal/view/J004136

Hickman, J. (2021). Developing and assessing students' spatial thinking skills when using GIS. *Teaching Geography*, *46*(3), 105–108. https://portal.geography.org.uk/journal/view/J003961

Holden, J. (2011). *Physical geography: The basics*. Routledge.

Hubbard, P., & Kitchen, R. (2011). *Key thinkers on space and place* (2nd ed.). SAGE.

Jo, I., & Bednarz, S. W. (2009). Evaluating geography textbook questions from a spatial perspective: Using concepts of space, tools of representation, and cognitive processes to evaluate spatiality. *Journal of Geography*, *108*(1), 4–13. https://doi.org/10.1080/00221340902758401

Jones, M. (2010). Limits to 'thinking space relationally'. *International Journal of Law in Context*, *6*(3), 243–255. https://doi.org/10.1017/S1744552310000145

Judd, N., & Klingberg, T. (2021). Training spatial cognition enhances mathematical learning in a randomized study of 17,000 children. *Nature Human Behaviour*, *5*, 1548–1554. https://doi.org/10.1038/s41562-021-01118-4

Kent, M. (2009). Space: Making room for space in physical geography. In N. J. Clifford, S. L. Holloway, S. P. Rice, & G. Valentine (Eds.), *Key concepts in geography* (2nd ed., pp. 97–118). SAGE.

Kobayashi, A. (2017). Spatiality. In D. Richardson, N. Castree, M. F. Goodchild, A. Kobayashi, W. Liu, & R. A. Marston (Eds.), *The international encyclopedia of geography*. John Wiley. https://onlinelibrary.wiley.com/doi/10.1002/9781118786352.wbieg1167

Kuebart, A., & Stabler, M. (2020). Infectious diseases as socio-spatial processes: The COVID-19 outbreak in Germany. *Tijdschrift voor Economische en Sociale Geografie, 111*(3), 482–496. https://doi.org/10.1111/tesg.12429

Laniado, D., Volkovich, Y., Scellato, D., Mascolo, C., & Kaltenbrunner, A. (2018). The impact of geographic distance on online social interactions. *Information Systems Frontiers, 20*, 1203–1218. https://doi.org/10.1007/s10796-017-9784-9

Larsen, T. B., & Solem, M. (2022). Conveying the applications and relevance of the powerful geography approach through humanitarian mapping. *The Geography Teacher, 19*(1), 43–49. https://doi.org/10.1080/19338341.2021.2008470

Law, M. (2022). Positioning geospatial: Classroom benefits and theoretical implementation. *Geographical Education, 35*, 24–32. https://agta.au/files/Geographical%20Education/2022/Geographical%20Education%20Vol%2035%20(2022)%20-%203.%20Law%20-%20Positioning%20Geospatial.pdf

Lefebvre, L. (1991). *The production of space* (translated by D. Nicholson-Smith). Blackwell.

Liben, L. S. (2017). Education for spatial thinking. In K. A. Renninger & I. E. Sigel (Eds.), *The handbook of child psychology* (6th ed.), Volume IV. Child Psychology in Practice (pp. 197–247). John Wiley.

Malpas, J. (2012). Putting space in place: Philosophical topography and relational geography. *Environment and Planning D: Society and Space, 30*, 226–242. https://doi.org/10.1068/d20810

Massey, D. (1991). A global sense of place. *Marxism Today, 38*, 24–29. https://banmarchive.org.uk/marxism-today/june-1991/a-global-sense-of-place/

Massey, D. (2005). *For space*. SAGE.

Morgan, K. (2004). The exaggerated death of geography: Learning, proximity and territorial innovation systems. *Journal of Economic Geography, 4*, 3–21. https://doi.org/10.1093/jeg/4.1.3

National Research Council. (2006). *Learning to think spatially*. National Academies Press. https://nap.nationalacademies.org/read/11019

Ness, D., Farenga, S. J., & Garofalo, S. G. (2017). *Spatial intelligence: Why it matters from birth through the lifespan*. Routledge.

Ortiz-Ospina, E., Roser, M., Ritchie, H., Spooner, F., & Gerber, M. (2022). *Migration*. Published online at OurWorldInData.org.

Richardson, L. (2022). Digital work: Where is the workplace and why does it matter. *Geography, 107*(2), 79–84. https://doi.org/10.1080/00167487.2022.2068839

Roser, M., Ritchie, H., & Ortiz-Ospina, E. (2015). *Internet*. Published online at OurWorldInData.org. https://ourworldindata.org/internet.

Sayer, A. (2000). *Realism and social science*. SAGE.

Scott, A. J., & Storper, M. (2015). The nature of cities: The scope and limits of urban theory. *International Journal of Urban and Regional Research, 39*, 1–15. https://doi.org/10.1111/1468-2427.12134

Sittner, T. (2021). A case for the curriculum: Health geography. *Teaching Geography, 46*(1), 21–24. https://portal.geography.org.uk/journal/view/J003827

Warf, B. (2019). Teaching digital divides. *Journal of Geography, 118*(2), 77–87. https://doi.org/10.1080/00221341.2018.1518990

Westlund, H. (2013). A brief history of time, space, and growth: Waldo Tobler's first law of geography revisited. *Annals of Regional Science, 51*, 917–924. https://doi.org/10.1007/s00168-013-0571-3

Chapter 6

Thinking geographically

Environment

As was stressed with the concept of place, the concept of environment needs to be distinguished from the environment as an object of study. The concept of environment is about how we conceptualise and understand the environment, and its significance to humans. It is a way of seeing the world that highlights the importance of the environment for human and other life, its influence on people and societies and the impacts on it of human actions. As a meta-concept, it contains a number of ideas, and these can be expressed as questions that geographers ask. These include:

- How do we define and conceptualise the environment, and the place of humans in it?
- How do we perceive and construct environmental phenomena?
- How does the environment support human life and wellbeing?
- How can we understand the environment?
- How does the environment influence people and societies?
- How are humans changing the environment?
- What are environmental problems, and how can they be explained?

These questions are all core themes in physical geography, as identified by Terence Day (2017),[1] but they build on and go beyond the typical content of physical geography in their exploration of the interrelationships between people and the environment. These interrelationships have been relatively neglected in geography until a few decades ago (Larsen et al., 2021), and some are still missing from school curriculums.

How do we define and conceptualise the environment?

The term 'environment' can have a variety of interpretations in geography, so it is important to explain what it means in this book. In the Macquarie Australian Dictionary, the word 'environment' is defined as 'the aggregate of surrounding things, conditions, or influences'. This could cover a very wide range of phenomena, but in geography, the term is usually restricted to the physical

DOI: 10.4324/9781003376668-7

and biological elements of the Earth. These include elements that we typically think of, but not always correctly, as 'natural', such as climate, landforms, rivers, soils, vegetation and animals. They also include things that have been produced by human actions, such as croplands, dams, parks, gardens, roads and buildings. This is an important point, because research shows that many students do not think of cities and agricultural areas as part of the environment. In this book, the environment described in this way may be called 'the biophysical environment' to make its meaning clear, or specific environmental types such as a marine, aquatic, terrestrial or alpine environment. The terms 'built environment' and 'urban environment' may also be used to refer to the constructed environments characteristic of the urbanised places in which the majority of students live, and which are a major site for geographical field work. Concepts such as the social environment, on the other hand, are better expressed as society and the political environment as politics.

The term 'natural environment' is sometimes used to describe much of the biophysical environment, yet if we think of natural environments as ones that have not been changed in some way by humans, then few such environments have existed for some time. As Jaboury Ghazoul writes in relation to the tropical forest:

> The concept of vast tracts of pristine tropical forest never exposed to human exploitation is now recognized to be mostly a myth. Almost wherever we look tropical forests show signs of having been affected to greater or lesser extents by a long history of human interventions, often limited only to hunting and occasional occupation, but sometimes extending to large-scale forest clearance to support urbanized civilizations.
>
> (Ghazoul, 2015, p. 66)

The term 'wilderness' has a similar problem. Western environmentalists think of wilderness landscapes as pristine, remote and unchanged environments that must be protected from human alteration, yet the Indigenous peoples who have lived in them for thousands of years regard them as their home and have managed and altered them over a long period of time. As the historical owners of the land, they also claim the right to develop its resources.

Today the extent of human alteration of environments is such that many scientists label the current geological era as the Anthropocene, in which the dominant influence on the environment is human activity, although note that the term has not been officially accepted as a new geological epoch. On the other hand, there are 'natural' phenomena and processes which exist and operate independently of humans, although they may be affected by human actions.

Are humans part of the biophysical environment? Research on children's perceptions reveal that many think they are not, while others accept that they are but still think they are distinct from it. In many cultures, humans are perceived as separate from nature, especially if this view is reinforced by a religious belief in humans as a special creation. However, because humans evolved from earlier primates, they are as much a part of the environment (or nature) as other animals. What differentiates them from related primates is a large brain, hands capable of manufacturing and using complex tools and an ability to communicate by voice which enabled them to develop new forms of social interaction and group organisation. With these attributes humans have developed complex cultures and ways of transmitting their knowledge to succeeding generations. However, these are not attributes that are unique to humans, as scientists are learning that some animals have quite complex ways of communicating through sounds, others such as bottle-nosed dolphins and chimpanzees can pass information to younger animals through teaching or imitation, and chimpanzees and an Indonesian cockatoo can make and select different tools for specific purposes. What distinguishes humans from other animals is that they have far greater powers to change the environment and to separate themselves from its direct influences. An approach advocated by the American physical geographers Michael Urban and Bruce Rhoads is therefore to view humans as embedded in and part of the environment, but having a unique agency in their ability to alter that environment while at the same time being influenced by it.

Landscape

The study of landscapes illustrates the ways that geographers perceive and construct the environment. Landscape is an important concept in geography, and at times has been proposed as the core of the subject. The most common meaning of the term is that it is the environment around us as seen through our eyes—the visual appearance of a distinctive assemblage of the material environment of landforms, vegetation, rivers, lakes, roads, buildings and other structures. A landscape may be entirely human-made, such as a formal park or garden, or an urban area, or it may be largely natural, such as a desert. Because they are defined areas, the term 'landscape' could be classified as belonging to the key concept of place, and the word has its origins in Germanic languages where it meant an area, territory or region (Huggett and Perkins, 2007). However, because landscape is about the biophysical and human features of the environment, and has stimulated integrated studies of the processes creating these features, I have classified it under the key concept of environment. Figure 6.1 shows a landscape, and below is a brief interpretation of what it reveals.

The Aroona Valley (the name is an Australian Aboriginal word for running water or place of frogs) has a semi-arid climate, with a median annual rainfall of about 210 mm and hot summers. The nearest line of hills in the scene is the ABC Range, produced by the erosion of a tilted bed of hard quartzite. Its European name is thought to relate to the number of its hills being equal

Figure 6.1 A landscape: a view of the Aroona Valley in the Flinders Ranges of South Australia.

Source: Author.

to the letters of the alphabet. The valley floor is soft siltstone. The vegetation of eucalypts and cypress pine (the latter so named by Europeans because it resembled northern hemisphere cypresses) is native to the area, but some may have been cleared for sheep grazing. The area is the traditional home of the Adnyamathanha People, who have lived there for at least 50,000 years and have changed the vegetation through their burning practices. They have their own explanations and names for the features of the landscape. Their lives and ownership of the land were disrupted when a sheep station was established in the valley in 1851, which further changed the environment, but the area is now part of a national park, the sheep are gone and the park was renamed Ikara-Flinders Ranges National Park in 2016 in recognition of its traditional owners. The photo was taken from Heysen's Viewpoint, a favourite spot of the painter Hans Heysen, whose paintings helped to change the way that Europeans viewed the South Australian landscape. The distant range in the photo is named after him. The Flinders Ranges are now an iconic Australian landscape and tourist destination. The scene reveals the influence of physical geography on the landscape; the effects of Aboriginal land management, European colonisation and government regulation; the belated recognition of Indigenous culture; and changing European perceptions of landscape.

Landscapes can be studied geographically in many ways. Physical geographers study their biophysical features, and specialise in specific ones, such as landforms or vegetation, or particular types, such as karst or alpine landscapes. They may collaborate with human geographers to explore the interacting physical and human processes that produce and change landscapes, using them as a place 'where the combined effects of society and nature become visible' (Bürgi et al., 2004, p. 857). Some human geographers study how humans perceive a landscape and the meanings it has for people. For example, some landscapes (such as the Lake District in England or the Grand Canyon in the USA) may be perceived as part of national identity and heritage, and in the UK, the term 'is commonly associated with a natural, rural and national heritage in need of vigilant preservation and protection' (Wylie, 2011, p. 305). Others investigate how the landscape was partly or wholly created by human actions and how different cultures produce visually different landscapes. Geographers also have been interested in how we perceive landscapes according to our culture, ideology and expectations. For example, the early British colonisers of Australia perceived the Australian landscape, and especially its vegetation, as drab and harsh compared with what they were familiar with.

Landscapes provide many opportunities for students to learn to perceive, question, investigate and interpret through fieldwork in a small area and to appreciate the interactions between biophysical and human processes.

How does the environment support human life and wellbeing?

The significance of the biophysical environment for humanity can be examined through the four ways it supports human life and contributes to human wellbeing. These are:

1 As a *source* of materials and energy.
2 As a *sink* into which humans dump wastes.
3 As a provider of ecosystem *services* that regulate and maintain the environmental conditions that enable humans to live.
4 As a *spiritual* influence on our emotions, health, imaginations and beliefs.

These could be called the four 'Ss'. The first of these will be familiar to teachers. It includes the production of food and materials from the natural resources of plants, water, soil, minerals and marine life, the provision of genetic resources and the power produced by fossil fuels, fissile materials and solar and wind energy. The second will also be familiar, as it includes the discharge of wastes into the atmosphere, rivers and lakes, soils, oceans and landfill sites.

The third category may be less familiar, as these services are not part of standard school physical geography. They include a wide range of processes that regulate and maintain the environmental conditions on which humans depend, such as:

- Atmospheric and ocean circulation patterns that redistribute heat and maintain stable climates.
- The hydrological cycle that circulates water between the Earth's surface and the atmosphere and makes life possible.
- The carbon cycle that, in the past, maintained stable atmospheric temperatures.
- The ozone layer that protects life on the surface of the Earth by absorbing ultraviolet radiation from the Sun.
- The ultraviolet rays in sunlight that kill parasitic, viral and bacterial pathogens in water.
- Soils, rivers and wetlands that break down, filter, store and dilute pollutants in water.
- Pollination by insects, birds, bats and wind, needed for most plants to reproduce.
- The processes that produce soils.
- The processes that recycle plant nutrients.
- The retention of rainfall by vegetation and soils, and its gradual release into rivers and lakes.
- Vegetation that reduces land and coastal erosion.
- Photosynthesis that releases oxygen from carbon dioxide and maintains the air we breathe.
- Predators that control pests.

These processes are commonly called ecosystem services; environmental processes that benefit humans without being managed or paid for and a rapidly growing field of theoretical and applied research. They operate without human involvement or even awareness, but can be damaged by human actions. For example, the ozone layer was not discovered until 1913, yet had been protecting humans from ultraviolet radiation for tens of thousands of years, and not until the 1970s was it discovered that it was being damaged by emissions of chlorofluorocarbons (CFCs).

The fourth way that the environment influences human life and wellbeing is through its influence on our emotions, health, imaginations and beliefs—in other words, on our spirit. These influences range from the feelings produced by a walk in a park or on a beach, or on viewing a beautiful or dramatic landscape, to the physical and mental health benefits of green or blue nature (Box 6. 1) and beliefs in the spirituality of the environment or the ritually purifying power of water. Some of these will be familiar to many students, but others may not. They remind us that the environment is much more than an economic resource to be exploited and that it sustains human life and wellbeing in more than utilitarian ways.

Box 6.1 Green space and health

A number of studies have investigated the relationships between the environment and people's mental and physical health and wellbeing. A review of this research covering seven countries concluded that the more green space in an area, the lower the mortality of the people in that area (Rojas-Rueda and Nieuwenhuijsen, 2019). The reasons suggested for this relationship were:

- Green space encourages physical activity.
- Trees and other vegetation reduce air pollution, noise and heat.
- Green space reduces stress and improves relaxation.
- Physical activity improves social connectedness.
- Exposure to a variety of microorganisms in green environments improves immune responses.

As another example, an innovative study in the UK collected geolocated responses from more than 20,000 participants on their sense of subjective wellbeing in different locations over time. These responses were compared with the land cover of each location, obtained from satellite images, and the researchers concluded that: 'On average, study participants are significantly and substantially happier outdoors in all green or natural habitat types than they are in urban environments' (MacKerron and Mourata, 2013, p. 992).

Underlying some of this research is the idea that humans have an inherent need for and connection with non-urban environments, because these are the environments in which we evolved and lived in until quite recently. Humans have lost much of that connection in a rapidly urbanising world, but the need remains and is the rationale for a range of projects to green urban areas and expose children to nature.

Combined, these four ways describe how the environment, thought of as the home of humankind, provides the conditions and resources that have permitted humans to survive physically and mentally, and for many but by no means all, at increasing standards of living. One consequence of these ways of describing the importance of the environment to people is that it can identify aspects of physical geography that students should understand if they are to appreciate the relationships between people and the environment. A view of geography as a study of the earth's surface is likely to emphasise climate, landforms and vegetation as the core topics in physical geography. On the other hand, a view of geography as a study of the relationships between people and environment might place more emphasis on understanding the influences described above and examine topics such as water resources, soils, pollution, ecosystem services, health, spirituality and urban planning.

Understanding the environment

Understanding the biophysical environment is the task of physical geography. In school geography, this typically involves studies of climate, geomorphology and ecosystems, or their integration in types of landscapes, such as arid lands, rivers and coasts. These are well described in existing texts and will not be discussed in this chapter. Instead I will focus on ways of thinking about physical geography that emphasise the complex interrelationships between the components of the biophysical environment, and between them and human activities, and which apply the key concept of interconnection. Research into students' scientific knowledge in several countries shows that many of them perceive the environment as an object, a collection of separate living things, and not as a web of interconnected biotic and non-biotic components and the processes that sustain them. Beliefs such as these are likely to affect students' understanding of environmental change, because they have a limited awareness of how the effects of human activities flow through these interconnections. To illustrate this way of viewing physical geography, this section explores ideas for teaching soils and rivers as topics that can link the different areas of the discipline.

Soil

Matthew Evans, in his book on soil, writes that 'soil has an image problem' (Evans, 2021, p. 1), and it tends to be a neglected topic in school geography curriculums and textbooks. The discussion of soil is mostly limited to its formation, types, physical and chemical properties and loss through erosion, often presented as separate material scattered through a textbook. Ian Selmes (2017) encourages teachers to view soil as the centre of an interconnected web with links to many areas of geography. He draws on a diagram by Janet Hudson, an amended version of which is in Figure 6.2.

Soil is the 'thick layer of biogeochemically altered rock or sediment at Earth's surface that has acquired numerous qualities during its exposure to the atmosphere that greatly distinguish it from its geological sources' (Amundson et al., 2015, p. 1261071-1). The characteristics of this layer of soil are influenced by climate (which affects weathering, leaching and organic matter decomposition), parent material (which affects weathering and the physical and chemical characteristics of the soil), relief (which affects erosion, weathering, drainage and exposure to solar energy), organisms (plants, animals and micro-organisms) and time. The combined influence of these five factors produce the soil-forming processes that in turn create the world's great variety of soils.

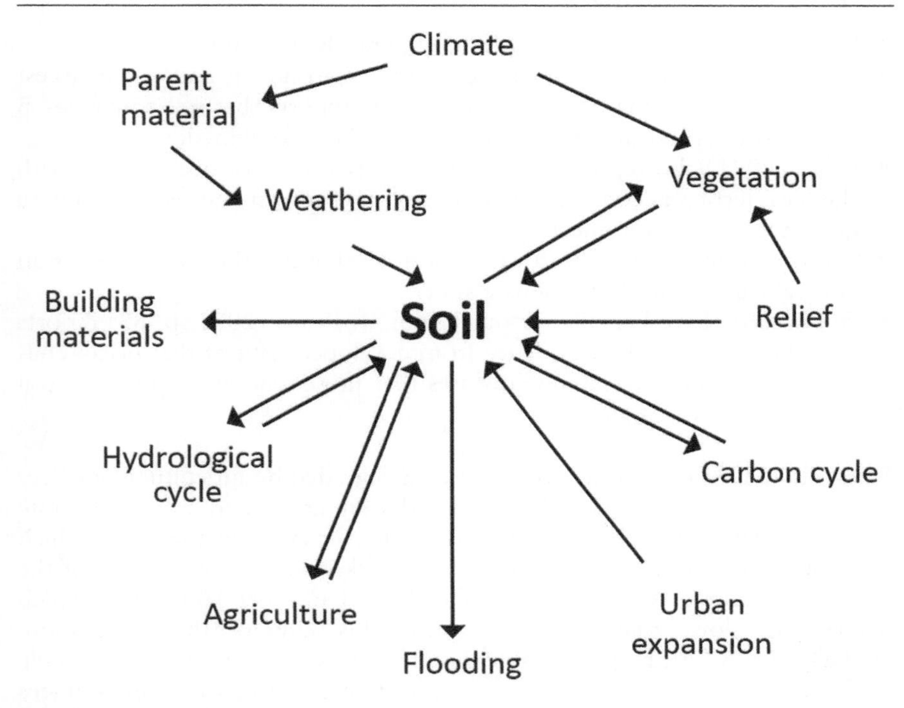

Figure 6.2 Soil's interconnections.
Source: Adapted by author from Selmes (2017, p. 95).

Why is soil so important? There are many reasons:

- Soil is the medium in which plants grow, so it supports all plant life, and plants in turn support animal and human life.
- The water stored in soil (called green water to distinguish it from the blue water in rivers, lakes and dams) supports 90% of the world's agricultural production and constitutes around 65% of the world's fresh water (Amundson et al., 2015, p. 1261071-1).
- Soil has pores that contain the air that provides the roots of plants with oxygen.
- Organisms in the soil help to decompose organic matter and make the nutrients in it available for plant growth.
- Increasing the organic matter in soils helps to retain more moisture.
- The ability of a soil to hold water has an influence on flooding, and when soils have been degraded by cultivation and erosion, the incidence of flooding may increase.
- Chemical, physical and biological processes within the soil filter and transform pollutants and improve the quality of water that flows into rivers and lakes.

- Soil is a source of materials such as clays, gravels and sands.
- Soil is a significant store of carbon in the environment and is the largest carbon pool after the oceans. Evans (2021) reports that soil contains 4.5 times more carbon than the rest of the biosphere combined.
- Evans (2021) also reports that healthy soil that has not been treated with chemical fertilisers contains microbes that breakdown and use methane (a greenhouse gas) from the air.
- Soil is a component of both the hydrological and carbon cycles, so is an integral part of the global climate system.
- Soil bacteria have benefits for our health, as Evans (2021, p. 53) reports that 'There is abundant evidence from developed nations that urban children have more autoimmune diseases that those who are exposed to soil bacteria'.

Soil is also a threatened resource. It is being degraded by agricultural practices that produce compaction, acidification and salinisation, and eroded at a rate faster than new soil can be formed. In Australia, for example, the rate at which new soil is produced by the weathering of rock is very slow, because of the low rainfall, and is estimated to average about 1 mm per 1000 years, which is very much slower than the rate at which soil is being lost through erosion. Globally, Evans (2021) quotes estimates of soil loss from the erosion of arable lands that range from 6 to 10 tonnes per person per year, a loss that reduces farm yields and food production.[2] Soils are also being degraded by the use of chemical fertilisers and herbicides that destroy soil biology. The alternative farming methods that could reduce these problems are a topic highly suitable for student investigation.

In studying soil in this way, students should learn that it is not only a vital and complex resource that supports human, animal and plant life, but also a threatened one, and needs to be protected.

Rivers

In schools rivers are mainly studied through fluvial geomorphology, focusing on the characteristics of a river—its gradient, flow, channel morphology and behaviour, and sediment transportation and deposition—and how these are controlled by the climate, topography, geology, vegetation, soils and land use of the areas through which it flows. However, we can also study the range of services that rivers and their associated wetlands and floodplains provide. The direct economic benefits of rivers include:

- Fertile floodplain soils, produced by the sediments deposited by rivers when they flood. These support much of the world's agricultural production.
- Accessible flat land suitable for buildings, industry and other physical infrastructure (see Lewin and O'Shea, 2023).

- Transportation by water, which was particularly important in connecting inland regions with the coast before the construction of railways.
- Water for irrigation and industry.
- Water for human consumption.
- Water landscapes for recreation and tourism.
- The removal of sewage.
- The generation of hydro-electricity.
- Food from fish and wild game.
- Improving water quality by breaking down, filtering, purifying, storing and diluting pollutants.

There are also indirect benefits. The wetlands associated with a river are important sites of high biodiversity and reservoirs of genetic diversity, but many have been drained, filled-in or converted. A review of 189 reports of wetland loss found that 57–61% of the world's inland wetlands have been lost since 1700 (Davidson, 2014). This represents not only a loss of biodiversity but also increases flooding, as wetlands absorb flood waters, slow water flow, store water for release back into the river and remove it through plant growth and evaporation. In addition, rivers and lakes are important places for recreation and the enjoyment of aquatic environments. There is also some evidence that regular visits to inland waters improve mental health in the same way that vegetation does.

Many rivers are regarded as sacred, and the Ganges is perhaps the most famous of these. Nick Middleton explains

> The water of the Ganges has numerous auspicious properties for Hindus. It acts as a medicine for every ailment, and bathing in it cleanses the devoted from all sin. Crucially, however, when a person's ashes or bones are entrusted to the river, the soul will be released for rebirth.
>
> (Middleton, 2012, pp. 38–39)

Indigenous peoples have strong connections to rivers and associated aquatic ecosystems, and these connections are often described simply as cultural. However, this misses the variety and complexity of Indigenous relationships with water. These include the supply of food through fishing, hunting and gathering; opportunities to develop tourism as a business; and the maintenance of river flows for places of spiritual significance. They also involve maintaining a living, healthy river system that people can care for and maintain customary relations with, because this contributes to their wellbeing. Leah Gibbs quotes the views of an Indigenous Senior Ranger on the significance of Dalhousie Springs, a large group of mound springs on the fringe of the Great Artesian Basin in northern South Australia:

> We have a holistic approach to water. For this water is a source of healing when we are sick, and it provides us with many spiritual and cultural

interests. It is our lifeblood, which we need to survive. It allows us to continue our ceremonies, which incorporate our rich and unique culture that is still strong today. It is these sources of water that provide an adequate and valuable food source rich in fish and other foods for my people.
(quoted from Ah Chee, 2002, in Gibbs, 2010, p. 374)

These uses of water are interconnected, and cannot be separated and managed individually, as is generally the case with Western methods of water management. The water that provides food is also the water that heals, so a holistic view is required. For example, for the Māori of Aotearoa New Zealand, 'an *awa* [river] is not just a river but an interconnected, living being that cannot simply be understood as a collection of measurable or definable parts' (Brierley et al., 2019, p. 1641). Rivers are much more than flowing water. In New Zealand, this Māori conceptualisation of a river has been recognised by the granting of the status of legal personhood to the Whanganui River in 2017. The legislation created a new legal entity, which described the river as 'an indivisible and living whole from the mountains to the sea, incorporating the Whanganui River and all of its physical and metaphysical elements' (quoted in Brierley et al., 2019, p. 1640).

Another application of the concept of interconnection to rivers is the way they connect the places and countries they flow through, which can be a source of both conflict and cooperation. For example, the storage and extraction of water by upstream countries or regions may threaten the supply of water to downstream places, as in the examples of the Nile River and the Tigris-Euphrates Rivers. Dam construction in Ethiopia is a potential threat to Nile River flows into Egypt, and dam construction in Turkey has created conflict with both Syria and Iraq over the sharing of Euphrates and Tigris water. Downstream places are also affected by pollution and sedimentation from upstream sources. So far, however, such disputes have been managed by negotiation and cooperation, and historically, serious conflict over water has been limited.

The discussion above has touched on only a few of the ways that geographers can study rivers. Others include:

- Human modifications of rivers. Many rivers have been modified by dams and locks, channel straightening, embankments, vegetation change, draining of swamps, extraction of water and pollution, and are far from 'natural'. Ashmore (2015) argues that such rivers should be studied as co-produced by geomorphological and social processes.[3]
- Floods and flooding, a popular topic in school geography.
- Stream and river restoration, which applies the skills of fluvial geomorphologists (Box 6.2).[4]

Box 6.2 An application of fluvial geomorphology

> As a result of the failure of a number of river management and engineering schemes there has been a radical shift in the nature of river management. This shift is towards working with, rather than against, natural processes and accepting the dynamic nature of river channels. This is because the long-term viability of many engineering structures cannot be assured given the highly mobile nature of rivers and the likely impact of a changing climate on river morphology in the future. Flood embankments can be lost to bank erosion, and current deflectors meant to deepen the river and curtail its lateral movement may well be largely ineffective in a river with high stream power. In this context, geomorphologists have a role in: (i) deciding on what type of rivers are activities such as channel straightening or floodplain development permissible; (ii) deciding where and how far from rivers structures can be built; and (iii) designing long-term and environmentally sensitive bank protection and engineering solutions to river management problems. Such an approach places the emphasis on living with rivers rather than fighting against the forces of nature.
>
> (Gilvear and Jeffries, 2017, p. 516)

The influence of the environment on societies

Does the environment have an influence on societies? Can it explain, at least partially, some aspects of the social world? To be more specific, do environmental characteristics such as climate, soils, vegetation, topography, water resources and rivers, have an influence? Do they explain some of the economic differences between places and countries? The answer is yes, but a qualified and contested yes. For example, climate, soils and water resources have a major influence on agriculture and pastoralism, and the patterns of human settlement they support, as do mineral resources on mining, and climate and topography on some types of tourism. These environmental influences are not direct, but depend on the culture, population density, type of economy, human organisation, level of technology and other factors of each society and on how they perceive, adapt to and use their environment. A low rainfall environment, for example, could be used for hunting and gathering, grazing of cattle for sale, growing crops with water supplied by irrigation, extracting minerals or natural gas, biodiversity conservation and nature tourism or the location of a city like Las Vegas.

However, there is a long history of thought that argues for a much deeper environmental influence on the nature of human societies, and a more recent one that argues that this influence is always mediated by human culture, institutions and technology. A brief review of this history can help in thinking about the interactions between humans and the environment.

Greek philosophers wrote about the influence of climate on the characteristics and development of societies, and later scholars speculated on the influence of topography and soils, in addition to climate, on human settlement, health, creativity, personality and economic activity. They argued that these features of society were strongly influenced, even determined, by the environment. This way of thinking became labelled as environmental determinism, and in the early 1900s was a focus for the relatively new academic discipline of geography in Europe and North America. A leading writer was the American geographer Ellsworth Huntington, who researched the effects of climate on people and societies. Another was Ellen Churchill Semple, who had studied in Germany under Friedrich Ratzel and whose influential book, *Influences of geographic environment*, was published in 1911. Their views were criticised by geographers and non-geographers as overly simplistic, historically and geographically inaccurate, uncritically based on often unreliable sources and unable to demonstrate that similar environments around the world produced the same social responses. Their ideas were also contradictory, with the effects of similar environments being described in opposite ways. As William Meyer and Dylan Guss (2017, p. 8) write, 'The effects of long, cold, and snowy winters, for example, were cited to account for both New England's supposed success and Russia's supposed backwardness'.

By the end of the 1930s, most Anglo-American geographers had rejected environmental determinism. Since the 1990s, however, there has been a growing number of publications, most of them by non-geographers, that renew the arguments for an influence of the environment on human history and economic development. This body of work has been labelled neo-environmental determinism by its critics, because it emphasises environmental factors over social ones and includes writers from biology, archaeology, history and economics, as well as journalists such as Robert Kaplan. Some their publications have been remarkably popular. They have also given many readers, both academic and non-academic, the belief that geography is the study of the influence of the environment on people and societies. Geographers, on the other hand, have been very critical of neo-environmental determinism, leading to the paradoxical situation of non-geographers arguing for geographical (which they define as environmental) influences and geographers arguing against, a reversal of their positions in the early 1900s.

An example of neo-environmental determinism is the argument of some economists that economic development in the tropics is hindered by disease and low agricultural productivity. The factual accuracy of this view has been strongly challenged, as per capita incomes and health standards vary considerably within each of the tropical and temperate zones, and Europe used to have high disease mortality until this was controlled by rising standards of living and investment in public health measures. Disease is more a result of poverty than a cause, as illustrated by the example of AIDS, which has 'become gradually more concentrated in lower-latitude countries for no climatic reasons whatsoever' (Meyer and Guss, 2017, p. 64).

In a recent study, Richard Tol, an economist, presents empirical evidence to show that the influence of climate on economic development is mediated by human institutions, has weakened over time and shrunk with affluence. His historical analysis of the spatial pattern of population density finds that the areas with the highest densities in 10,000BCE had a temperature of 23°C and 3000 mm/year rainfall. He suggests that 'Early humans liked it warm and wet. With primitive heating and clothing, cold winters were deadly. High net primary production in the tropics meant abundant food'. However, in 100CE, he identifies strong population growth in temperate climates and suggests that 'As agriculture, irrigation, clothing, heating, and cooling improved, humans moved into hitherto inhospitable areas' (Tol, 2021, pp. 18 and 19).

Furthermore, studies of the influence of climate on economic development generally ignore the historical interconnection between the temperate countries and the tropical world. The effects of this relationship were detailed in Clifford Geertz's (1963) classic study of the development of the economy of Java in Indonesia. He argued that in the early 1800s, Java (tropical) had an economy quite similar to Japan (temperate), with relatively dense populations dependent on labour-intensive rice cultivation on small farms, but between 1830 and 1914 their development paths diverged. Japan eventually built a manufacturing sector which absorbed much of the growth in the labour force and raised incomes, while in Java much of the population growth was absorbed into ever more intensive but low-income farming. Japan restructured and urbanised its economy, while Java did not. Geertz argues that the main reason for the divergence was that Java was a colony of the Netherlands. This meant that the wealth produced in Java through crops grown for export financed the economic development of the Netherlands, not Java.

These debates may seem irrelevant to school geography, but the influence of writers such as Robert Kaplan and Jared Diamond has ensured that neo-environmental determinist still exists in popular opinion and may influence students.

The concept of materiality, used by cultural geographers and related social scientists, offers another way of thinking about the influence of the environment on societies. Materiality refers to the specific material properties of something that can have social and political outcomes. Water is a good example. It is a very complex material, because it exists in and flows through the environment in different forms, is very variable over space and time, is essential to all life, has many competing commercial uses, as well as recreational, aesthetic and spiritual values, and can be degraded by pollution. This combination of properties make the management of this resource particularly difficult, and has led to the development of a variety of institutions to control and distribute water, from the *subak* system of Bali, which evolved centuries ago to manage the supply of water for rice cultivation, to new forms of regional governance in California to manage interconnected water problems early in the twentieth century. In both examples important social institutions have been created to deal with the specific properties of water.

At present many geographers seem reluctant to admit any significant influence of the environment on human societies, perhaps because of the past excesses of early environmental determinism. Meyer and Guss (2017, p. 97) argue that some geographical critiques of neo-determinism err 'on the opposite side by disregarding the important role often—though not independently—played by the environment'. This role is not independent because it is mediated by economic, political and cultural circumstances that change over time and space. Natural hazards provide an example. Different environments have different natural hazards, but their impact on people is determined by human as well as environmental factors. For example, when housing is allowed on floodplains the damage produced by a flood will be increased; when sand dunes are covered with buildings and roads the damage to beaches from storms is also likely to be increased; and the drainage of wetlands may increase the damage from flooding by removing areas that once absorbed flood waters. Humans are also having an influence on natural hazards as global warming is increasing the frequency and/or severity of hazard events. Natural hazards are not purely 'natural'.[5]

The influence of people on the environment

People have been changing the biophysical environment for the whole of human history, and an understanding of many features of present environments requires some knowledge of their past. Australia provides a good example. The early European explorers and colonists observed that many areas were not thickly forested, with one district in South Australia being described as 'partly wooded, partly clear [with] more the appearance of an immense park than anything that one would naturally expect to find in the wilds of an uncultivated land' (Gammage, 2011, p. 16). The landscape described in the quotation was produced by Aboriginal burning practices. In many parts of Australia, carefully managed and low-intensity burning was used to increase the productivity of food gathering and hunting. This involved burning to produce new grass to attract game, which was done in patches, leaving areas of dense vegetation for animals to shelter. Fire was also used to drive game to where they could be caught, clear land for harvesting, regenerate useful plants and make travel easier. The general outcome of these practices across non-arid Australia was an increase in grassland, a reduction in forest, a reduction in the understorey of wooded areas and the creation of a patchwork or mosaic of vegetation types. The landscape that Europeans observed and admired was created and managed by Aboriginal people, whose knowledge and skills maintained a productive environment for both humans and other lifeforms.

In other parts of the world, people also changed environments to grow crops, pasture stock, build villages and towns and produce minerals, but by the 1800s, economic growth and the resulting exploitation of the environment were creating significant damage. This was first extensively described for English speakers by the American George Perkins Marsh (1801–1882) in his book *Man and nature, or, physical geography as modified by human action*, published in 1864 (the book was written at a time when 'man' was an acceptable word for all humans).[6] The book has been described as 'the first great work of synthesis in the modern period to examine in detail man's alteration of the face of the globe' (Thomas, 1956, p. xxix), and it warned Americans of the consequences of the damage being done to their environment. Marsh is particularly remembered for this much-quoted statement; 'But man is everywhere a disturbing agent. Wherever he plants his foot, the harmonies of nature are turned to discords' (Marsh, 2003, p. 36). Marsh, however, had been strongly influenced by Alexander von Humboldt, whose early thoughts on the effects of humans on the environment were based on the effects of forest clearance by pre-European societies on soils, water resources and climate that he observed in his travels through Central and South America at the beginning of the 1800s. Humboldt's books were widely read in America and Europe, and Marsh called him 'the great apostle' (Wulf, 2015).

A later influential book was *Man's role in changing the face of the earth* (Thomas, 1956), which was the result of an interdisciplinary and international symposium held in 1955. Fittingly, it was dedicated to the memory of Marsh and recorded almost everything that was then known about human alteration of the environment. A current and more accessible book on the same themes is British geographer Andrew Goudie's *Human impact on the natural environment*, now in its eighth edition (Goudie, 2019). It has a brief introduction to the development of thinking about human transformation of the environment, and teachers could use it to give students some awareness of the history of environmental degradation. They will learn that the issues they may be concerned about are not new, with the important exception of global warming.

An awareness of the changes made by humans to the environment should lead students, when studying a landscape or an environmental issue, to think of whether any of the following apply to it:

- Clearance of the original vegetation, which may have caused soil erosion and/or soil salinity, and possibly landslides or slumping. It may also cause changes in local climates and reduce rainfall.
- Clearance of coastal vegetation, such as mangroves, which may have increased storm damage.
- The effects of atmospheric pollution on vegetation.
- Planting of new vegetation (such as pastures, crops and forests).
- Introduction of exotic animals, which may have resulted in land degradation.

- Agriculture and grazing, which may have caused soil degradation and erosion.
- Drainage of wetlands and their replacement with farmland, which may have increased flooding.
- Regulation of rivers and modifications to river channels.
- Dams that created lakes and changed downstream river flows.
- Construction of railways and roads, and their effects on water flows and erosion.
- Coastal structures that changed the movement of sediments.
- Increased downstream sedimentation caused by upstream accelerated erosion.
- Mining.
- Urban development (see later in this chapter).

This list is by no means complete, but it may help to remind teachers of the variety of human impacts on the environment. The consequences of this for physical geography are described by Michael Urban

> In our own time, the rapid rate of environmental change coupled with the sheer magnitude of human drivers have propelled geographers to increasingly examine landscapes that are as much a byproduct of social policy or individual action as the laws of physics, chemistry, or genetics.
>
> (Urban, 2018, p. 49)

He therefore suggests that geographers should be focussing on what he calls 'crappy' landscapes. These are the greatly altered landscapes in which people live, such as urban or farming areas, rather than the glaciers, coasts and rainforests which are often the focus in school geography (see also Lane, 2019).

Human alteration of the environment is not a simple one-way impact, because the environment may respond in ways that require further human actions. For example, a study of stream channelisation in a watershed in western Tennessee [USA] found that:

> The initial straightening and dredging, undertaken to reduce flooding and assist agriculture, thereby set off a series of human-landscape feedbacks in which streams adjusted to the higher energy of steeper gradients by eroding their banks and beds, which increased sediment loads, which then led the landowners to re-dredge.
>
> (Harden, 2014, p. 80)

This is an example of feedback, an important concept that was noted in Chapter 3. It also illustrates the deficiencies of engineering solutions to river problems, a point that will be returned to at the end of the chapter.

Explaining environmental problems

Environmental problems are environmental changes that have negative impacts on people and other life forms. They are not just 'environmental' problems but also social and economic ones. They include:

- Loss of biodiversity in animal and plant life which in turn threatens the productivity and resilience of ecosystems.
- Land degradation that reduces agricultural and pastoral production and threatens livelihoods.
- Ocean overfishing, pollution and acidification that threaten food supplies.
- Atmospheric and land pollution that affects human and ecosystem health.
- Vegetation clearance that contributes to land degradation, global warming and reductions in rainfall (examined in Chapter 10).
- Coastal erosion that threatens homes, buildings and infrastructure.
- Global warming that threatens food production, labour productivity and human health.
- Unsustainable extraction of water from rivers, groundwater and aquifers.

All of these are the result of human actions interacting with environmental conditions and biophysical processes. However, while they are current issues, there is nothing new about humans changing the environment. The early hunter-gatherer societies that originated in Africa and spread to other continents were long assumed to have had no impact on the environment, because of their apparently simple technology, but have now been shown to have altered the vegetation by sophisticated methods of burning designed to manage game and increase food production. The later agricultural societies that separately developed in several regions of the world permanently cleared land for farming, and many developed complex methods of irrigation that changed the local and regional hydrology, and in some areas degraded soils through salinisation. Another consequence of agriculture was that the levels of greenhouse gases in the atmosphere started to increase more than 3000 years ago as a result of the expansion of agriculture and the clearing of forests. More recently, the Industrial Revolution increased environmental change through mining, commercial agriculture, industrialisation, fossil fuel use and urbanisation, and these changes have accelerated since around 1950. Population growth has also added to the pressures on the environment.

Many of these problems are well covered in school textbooks, but what is sometimes missing is an adequate investigation of their causes. Understanding causation is important because identifying ways to respond to an environmental problem depends on an accurate understanding of what is causing it. This can be complex, because the causes are both environmental and social, and can operate at several levels of scale, as discussed in Chapter 2. Ways of thinking about the causes of an environmental problem are illustrated below through case studies of urban environments, land degradation and climate change.

Urban environments

Urban environments are one of the most interesting ones for school study, because they are the environment in which the majority of students live, and illustrate very clearly the effects of human alteration. Changes include replacing vegetated surfaces with constructed impervious ones, such as buildings and roads, altering the hydrology through constructed drainage and water supply systems and modifying the vegetation cover. The result is a different environment, particularly in its climate, hydrology and biodiversity, and some of these changes have a negative impact on human health and wellbeing as well as on the environment. These are described and explained below.

Cities are often hotter than non-urban areas, a phenomenon called the 'urban heat island' effect, where temperatures in central city areas are higher by from 4°C to 6°C or more than in the surrounding suburbs and rural areas. Figure 6.3 maps the rise in temperatures towards the centre of London for a typical day in May. At a continental scale another study estimated that, on

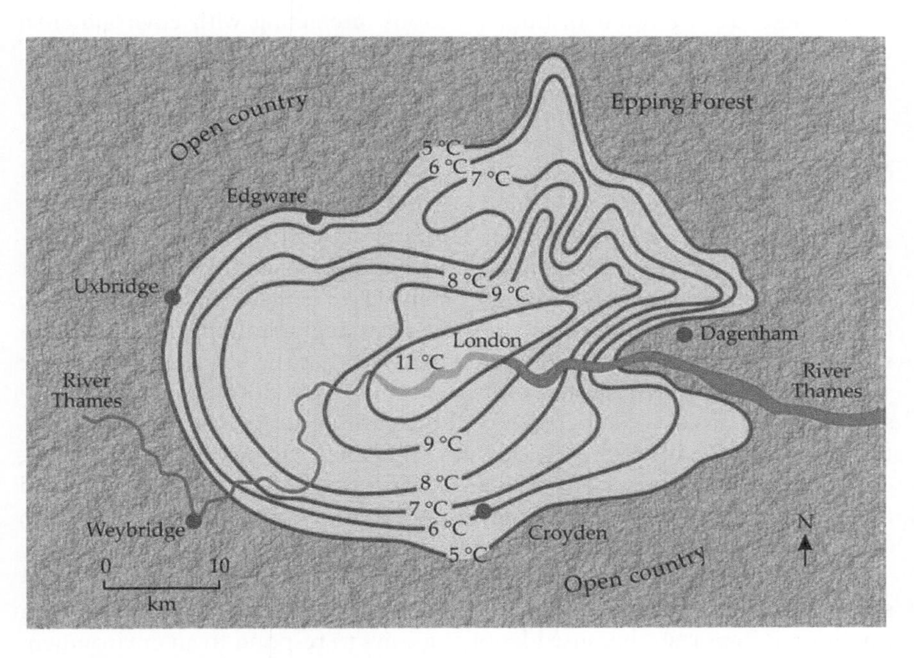

Figure 6.3 The urban heat island effect in London.

The map shows the minimum temperature on a typical day in May rising towards the centre of the city.

Source: Reproduced from Pielke et al. (2016, p. 44) with the permission of AIP.

average, US cities are warmer than vegetated lands by 1.9°C in summer and 1.5°C in winter (Bounoua, 2015). However, cities built in arid environments, where irrigation systems support vegetation with high levels of transpiration, are cooler than the surrounding areas.

Urban heat islands have several causes:

- Dark surfaces like roads, roofs and car parks absorb more of the sun's heat than grass and trees (i.e., they have a low albedo), and buildings and other artificial materials store heat in the materials used in their construction. These materials also release heat more slowly than trees and grass, so the urban atmosphere cools more slowly at night than in the surrounding rural areas.
- The energy used for heating, cooling, production and transportation creates heat, as do human bodies.
- The removal of vegetation for urban development reduces the cooling effect of the shade and transpiration provided by trees. However, note the comment above about the cooling effects of irrigation in cities in arid areas.

Higher city temperatures are an environmental problem because excessive heat is a significant cause of human mortality and morbidity and may also be an economic cost because of the increased need for cooling. The heat island effect is also unequally experienced within cities, as lower-income suburbs often have a poorer provision of green spaces, which have a cooling effect, than higher-income suburbs. This makes their residents more vulnerable to the heat waves that are likely to become more frequent with global warming.

Cities are drier than non-urban areas, because their impervious surfaces prevent water from entering the soil and constructed drains rapidly remove water to rivers and coastal waters. This is a problem because drier environments have less cooling from evapotranspiration.

Cities are more prone to flooding after heavy rainfall, because surface runoff is faster from impervious surfaces than from vegetated land, and the wetlands that once absorbed flood waters may have been filled in and built over. Furthermore, the heat of the city produces convection, and the upward air movement increases the frequency of thunderstorms. It also draws air in from the surrounding rural areas, which is likely to be moister and more conducive to rain.

The land surface of some cities is sinking because of the extraction of groundwater, which leads to increased flooding and structural damage to buildings and roads. Jakarta in Indonesia is an example of this problem, where studies estimate that areas of the city are sinking between 1.8 cm and 10.7 cm per year. The main cause of this subsidence is the massive use of groundwater by residents and businesses, a practice that regulations have failed to adequately control. When combined with slow sea level rise as a result of climate change, the result is widespread flooding (Aldrian, 2021). Other major cities that are sinking because of excessive groundwater extraction include Houston,

New Orleans, Mexico City, Bangkok, Ho Chi Minh City, Tokyo, Venice and Rotterdam.

Pollution is a common urban environmental problem. Soils can become polluted with waste materials and industrial discharges, some of them toxic. Water ways can be polluted from the fertilisers and pesticides used in gardens, parks and road verges and from metals and hydrocarbons washed off road surfaces during rain events of sufficient magnitude. Some of these may be in toxic concentrations. Air is polluted by gases and particulate matter produced by vehicles, energy generation and industrial processes and levels of atmospheric pollution in many large cities are regularly above accepted health limits. Studies again show that the most socioeconomically deprived urban areas tend to have the worst air quality.

From this understanding of the causes of changes to the environment produced by urbanisation, students should be able to suggest ways to respond. Table 6.1 lists causes in the left-hand column and the corresponding responses in the right-hand column, to show how they are related. All of these responses have been implemented in different places around the world.

Table 6.1 Causes of and responses to the effects of environmental change in urban areas

Cause	*Responses*
Dark surfaces	Covering roofs with highly reflective materials that do not absorb as much heat. This will also reduce the energy used for cooling buildings.
Impervious surfaces	Using permeable materials for paths, driveways and car parking areas that allow water to pass through into the soil. Using pervious asphalt for neighbourhood roads. Constructing soft verges for roads to absorb water.
Heating from burning of fuels	Reducing energy use through improving energy efficiency, and reducing the use of fossil fuels.
Lack of green vegetation	Creating green roofs with shrubs, grasses and even trees. The vegetation will absorb less heat, and its evapotranspiration will cool the air. Planting more trees along roads and in parks and gardens, and protecting existing ones. Clustering medium-density housing to retain land as communal green space and animal and plant habitat.
Constructed drainage systems that remove rainwater from the environment	Diverting rainwater from house roofs back into gardens where these are large enough to absorb the water. Diverting stormwater on roads into kerbside soakage pits, or local detention ponds and wetlands. These will also trap and break down pollutants in the water. Restore some of the pre-urban features of rivers and streams.
Groundwater extraction	Providing alternative sources of water, including recycled grey water. Diverting stormwater into groundwater aquifers to replenish them.

There are several things that students could learn from a study of urban environments.

- How to use an analysis of causes to identify responses to problems.
- A deeper understanding of environmental processes. For example, understanding why changing the hydrology of cities produces problems leads to a better understanding of the functioning of less altered streams and rivers, and why river restoration is increasingly supported by government agencies and community groups.
- How the impact of environmental problems is often unequally distributed across space, with places with socioeconomically disadvantaged groups more likely to be affected than places with more advantaged populations. This adds an important social justice dimension to the study of environmental problems.
- The concept of restoring 'natural' environmental processes to manage problems. All the strategies listed in Table 6.1 are designed to at least partially restore the ways that the environment functioned before urbanisation. This is an example of 'nature-based solutions' to environmental problems, a concept that is discussed later in this chapter.

Land degradation

Land degradation is a decline in the quality of soils and vegetation that reduces their ability to produce food and support livelihoods. It can take many forms, such as:

- Soil erosion.
- Vegetation degradation, which reduces feed for livestock and wood for fuel and construction.
- Soil salinisation.
- Other forms of soil degradation, such as acidification, compaction and loss of soil organic carbon.

Land degradation is a problem in all parts of the world. The European colonists who occupied and farmed large areas of North America and Australia created widespread erosion by using inappropriate farming methods in an unfamiliar environment often prone to drought. They also cleared large areas of their tree cover, which degraded the hydrology and biodiversity of these areas. While the magnitude of degradation is probably greatest in high-income countries, the rate of degradation in these countries has slowed, whereas the region currently most affected by increasing land degradation is Sub-Saharan Africa and is the focus of this section. The discussion is organised as a series of questions and answers that might encourage students to see that there can be different viewpoints about an environmental problem.

How accurate is our knowledge of the extent and severity of land degradation in Sub-Saharan Africa? Measures of land degradation were initially based on the opinions of experts in different countries and regions, rather than on any objective measurement. More recently, remote sensing been used to measure some environmental variables, such as land cover and net primary productivity. This method has the advantage of uniform coverage of large areas but has several limitations. One is that so far it can only measure a few variables, and much of the information needed for policies to respond to land degradation have to be obtained by local field surveys. Another is that different measures produce significantly different maps of land degradation, with limited agreement between them, and scientists argue that it is only possible to measure particular types of land degradation and not some overall concept of degradation. A third problem is that to measure degradation, it is necessary to measure change, which requires comparable data for several decades and this information is scarce in Africa. Nevertheless, estimates of land degradation exist, such as one by David Ussiri and Rattan Lal (2019, p. 2), who write that: 'Consistent studies covering wide geographical areas of SSA [Sub-Saharan Africa] are limited, but some studies suggest that about 500 Mha of Africa (about 16% of the continental land area) are affected by different types of degradation'. An Intergovernmental Panel on Climate Change (IPCC) report, on the other hand, reports that land degradation occurs over a quarter of the Earth's ice-free land area and affects 1.3–3.2 billion people, the majority of whom live in poverty (Olsson et al., 2019, p. 347).

Other research studies question the extent of land degradation, arguing that it is highly variable within quite small areas and difficult to generalise over larger areas. Furthermore, it is contended that land degradation has become a 'narrative' adopted by international environmental agencies to gain attention and funding, as Roy Behnke and Michael Mortimore explain:

> If scientists require clarity in the concepts they employ, the politicians and administrators who create and manage large institutions have other, very pragmatic requirements. In the search for money and support, they need a problem that is dramatic enough to command immediate attention, simple enough to be quickly grasped, and general enough to satisfy diverse interest groups; they need what Jeremy Swift has called a development narrative—a powerful story line with clear, broadly applicable policy implications and urgent funding needs.
>
> (Behnke and Mortimore, 2016, pp. 5–6)

Over time, this narrative about land degradation became simplified and increasingly disconnected from the scientific evidence.

Is land degradation solely caused by human actions such as poor farming methods, overstocking of cattle and deforestation? Lawrence Kiage, a physical geographer, argues that:

> In most cases it takes interaction between human activities and the biophysical environment for soil erosion and land degradation to occur .... However, most literature and reports have tended to exaggerate anthropogenic impacts relative to the biophysical predisposition of the land. Biophysical factors including soils, climate, topography, and vegetation cover have important roles in the soil erosion process.
>
> (Kiage, 2013, p. 666)

It is this interaction between human activities and the biophysical environment that should make land degradation of particular interest to geographers. Furthermore, to understand the biophysical factors that Kiage lists, students will need to understand the effects of soil materials, rainfall, slope and vegetation on soil erosion, which they may be able to observe for themselves not far from their school.

What are the underlying causes of the human actions that are degrading land? Population growth is frequently blamed for increasing the pressures on land that cause land degradation, as explained in this account:

> Such rapid population growth leads to intensified stresses on natural resources—water, land, forest, and pasture in rangeland ecosystems. The demand for cropland expansion leads to expansion into more marginal conditions unsuitable for agriculture. The stress on natural resources and ecosystems often translates into overexploitation through overcultivation, deforestation, and overgrazing, and also poor land management practices. The consequences of stresses on ecosystems are soil erosion and degradation that perpetuate food insecurity and poverty in SSA.
>
> (Ussiri and Lal, 2019, p. 8)

Kiage, however, while not disagreeing with this assessment, argues that the linkages between population growth and land degradation are 'neither clear-cut nor direct', and he notes research in Sub-Saharan Africa that found that population growth led to the adoption of improved land management practices and a reduction in soil erosion. Often overlooked in studies of land degradation is the way that local people may respond and adapt, and the circumstances in which they can do this.

Overstocking is another commonly listed cause of land degradation. Kiage again cautions that:

> A number of scientists suggest that grassland communities are not normally damaged by grazing. Animal mortality usually causes a fall in numbers that coupled with migration cushions land from damage. The controlling factor is actually the amount of dry season vegetation. When the rains return, usually insufficient animals remain to threaten recovery of the rangelands.
>
> (Kiage, 2013, p. 672)

Communal land ownership has also been blamed for land degradation, because it is argued that where land is owned by a group rather than individually, there is no incentive for people to limit their use of common land. This situation is called the 'tragedy of the commons' and is often mentioned in school textbooks. Yet a number of research studies have shown that this was 'neither common nor inevitable', by demonstrating that 'self-organized collective institutions governed by stable communities that are buffered from outside forces have mostly sustained common-pool environmental goods and services successfully' (Montanarella, Scholes and Brainich, 2018). Kiage adds that communal land ownership in Sub-Saharan Africa was not a free-for-all, because:

> Elaborate institutional arrangements, rules and regulations existed that governed land use and ensured sustainable utilization of resources by members. Soil erosion and degradation, while mistakenly attributed to common property systems, actually began with the dissolution of local-level institutional arrangements whose very purpose was to give rise to sustainable resource use patterns.
>
> (Kiage, 2013, p. 674)

Finally, poverty is a commonly identified cause of land degradation, as in this extract from a report on Sub-Saharan Africa: 'Poverty is a driver of soil degradation when farmers and herders resort to inappropriate land management practices such as farming in marginal lands, overgrazing, overstocking in marginal grazing lands, and the elimination of fallow periods' (Ussiri and Lal, 2019, pp. 8–9). On the other hand, Samantha Jones notes a number of studies that found that the poorest do not necessarily degrade the land the most:

> Moseley (2005) found that in Mali the soil quality measures practiced on the farms of rich and poor households are not significantly different and that the wealthy pursue more soil deleterious practices through cotton export-orientated production. Gray (2005) found that in Burkina Faso, wealthier farmers have fewer trees and higher levels of animal traction that destroys soils compared to poorer farmers, while poorer farmers tend to conserve environmental resources at the expense of their own well-being. These studies tend to stress the adaptability and agency of the poor.
>
> (Jones, 2008, p. 679)

In the extracts above from Kiage, a geographer originally from Kenya, he questions and qualifies several of the commonly accepted explanations for land degradation and provides an important African perspective on the topic. In the article from which these extracts are taken, he also criticises external development agencies for not recognising and supporting the adaptation and coping strategies employed by local people.[7]

Is there another level of explanation for land degradation? The causes of land degradation described above operate at local and regional scales, but there are other causes that operate at national and global scales. Examples include:

- Increasing national and global demand for livestock products resulting from population growth and rising incomes, which in turn is responsible for an expansion of cropland to produce animal feed.
- Increasing global demand for crops such as sugar, oils, coffee and cacao, which increases the pressure on land resources.
- Government policies that favour export agriculture, combined with insecure land rights, and poor governance and legislation, which in countries such as Kenya have resulted in many small- and medium-sized farmers losing their land to large producers, so putting greater pressure on the remaining land resources for food production.
- Economic globalisation, whose effects are described in a major international report as

> in a world increasingly interconnected economically and ecologically, the costs imposed by environmental and land degradation are felt disproportionately by low-income nations which are being increasingly depended upon as the producers of raw materials for more developed nations.
>
> (Montanarella et al., 2018, p. 190)

Students could learn the following from a study of land degradation:

- To be cautious in accepting popular environmental narratives such as land degradation and desertification.
- To be aware of how perceptions of an environmental problem may differ between local people and outsiders.
- To critically examine common explanations of an environmental problem, such as population growth or poverty.
- To look for explanations beyond the immediate actions of local people to the wider processes that influence these actions, including unequal power relations between and within countries.
- To investigate both biophysical and social explanations.
- To recognise that our knowledge of environmental problems is continually evolving.

The discussion above illustrates an approach to the integration of physical and human geography that has long been advocated by geographers, but seldom achieved. Physical geographers have generally included limited and unsophisticated analyses of human influences, if they are included at all, while human geographers have rarely treated the environment 'as anything more than a space in which human activity occurs' (Lane, 2019, p. 50). Effective integration requires a deep analysis of both the biophysical and social processes that have co-produced the situation being studied, as well as the use of both qualitative and qualitative methods.[8] Sheppard argues that:

> For geographical thinking to approach its potential, those from the human side need to familiarize themselves with the role of biophysical processes, whereas those from the physical side need to familiarize themselves with how societal processes (both political-economic and cultural-representational-performative) are integral to the biophysical phenomena they study.
> (Sheppard, 2022, p. 19)

Climate change

Climate change is a difficult topic for teachers because it is complex, multidisciplinary, controversial, affects the lives of everyone and is a worry for many young people. It also has an impact on many of the topics studied in school geography, from vegetation to human migration. This section discusses what geography teachers could contribute to teaching about climate change.

To start with, they should be familiar with the reports produced by the IPCC at https://www.ipcc.ch/reports/, as these are the most authoritative sources of current knowledge, and should be preferred over media and non-scientific interpretations. The Sixth Assessment (2022) has four reports: a synthesis and three more detailed thematic reports. The synthesis report may be of most value to busy teachers, as it summarises both the other three and several additional IPCC special reports. The thematic reports have lengthy summaries, which will help to identify the relevant sections in the body of each report. Two of them also have an atlas version, which can be used to illustrate spatial analysis.

Geography teachers should ensure that students understand the science of global warming, but as they are not climate scientists, this should be taught in science and drawn on in geography.[9] Students should know what greenhouse gases are and why they produce global warming. They should also know how scientists can determine that these gases are warming the atmosphere, which means some understanding of the modelling that identifies the effect of greenhouse gases on atmospheric temperature. This modelling compares simulations of past temperatures that only include natural factors such as solar, volcanic and ocean-atmosphere variability, with those that add the human influences of greenhouse gas emissions, ozone and land use changes. Only the second simulation is able to produce modelled temperatures that match the observed ones (an example of this modelling is shown in Figure 7.1). The IPCC is now unequivocal that human influence has warmed the atmosphere, ocean and land.

Students should also know some of the other indicators that show climate is changing. The ones below come from IPCC reports:

- The global retreat of glaciers.
- A decrease in Arctic sea ice.
- Sea level rise of 3.7 mm. a year between 2006 and 2018.
- More frequent and more intense heat extremes.
- More frequent and more intense heavy rainfall events.
- The many species in terrestrial, freshwater and marine ecosystems that have shifted their range towards higher latitudes (i.e., away from the equator).
- Coral bleaching and mortality.
- The poleward movement of pests and diseases.

Specifically geographical teaching about climate change could include the following.

1 How greenhouse gas emissions vary from place to place? This is fundamental to understanding the negotiations between higher-income and lower-income countries over who is responsible for global warming and how de-carbonisation should be funded. It also helps students to understand how countries vary in the changes they will have to make to reach net zero emissions.
2 Which regions of the world are projected to be the most affected by future climate change, and how can they adapt? The spatial pattern of the effects of climate change on temperatures, rainfall and sea level rise is likely to be very uneven, as is the capacity of communities to adapt, and this is an important area for geographical research.
3 How are specific environments and landscapes being affected by climate change, and how are they responding? Geographers' interest in place provides a useful corrective to generalising global narratives, with local studies that examine how global processes interact with the characteristics of individual environments to produce different outcomes. Geographers also study how environments respond to climate change and the resilience with which they adapt.
4 How might global warming affect:
    - migration?
    - food production?
    - natural hazards?
    - cities?
    - coastal regions?
    - human health?
    - Indigenous people around the world?

There is material on all these topics in the IPCC reports, findable if you use the indexes.[10]

5 Can a country claim to have reduced its greenhouse gas emissions if some of the reduction has been achieved by shifting production to other countries? Carbon dioxide is embodied in the commodities imported by countries, and studies find that the apparent success of the USA, Japan and much of Europe in reducing domestic carbon dioxide emissions has been achieved by outsourcing production to other countries (Kanemoto, Moran, Lenzen, and Geschke, 2014). This is an application of the key concept of interconnection.
6 What is the role of land cover change in climate change? Land cover change is a longstanding theme in geography and in recent years has been increasingly recognised as a cause of climate change. This theme is discussed in some detail in Chapter 10.
7 How can we reduce greenhouse gas emissions to net zero? To answer this question, students must understand the causes of these emissions, so that the strategies they choose are ones that will address them. The paragraph below outlines an approach to understanding the causes through the idea of layers of explanation.[11]

Carbon dioxide is the main greenhouse gas and is produced by the actions of people when they use energy from fossil fuels in transport, heating, cooling and appliances, and by industries when they use them to operate machinery, make products and transport materials. So is the problem caused by the behaviour of people and businesses, which is the first layer of explanation? As explained in Chapter 8, this is what ecological footprints measure, and use to encourage people to change their behaviour, yet the ability of people and businesses to change is constrained by the choices available. This is then the second layer of explanation, as until recently consumers had no ability to choose electricity produced from renewable sources, or to drive electric vehicles, and the recent rapid take-up of solar panels in some countries suggests that many people do want an alternative source of energy. So why is so much electricity still generated from fossil fuels, which is the third layer of explanation? Is it because the technology is not yet available to replace coal and gas, or are the alternatives too expensive? I don't think either of these is true, and Michael Mann in his book, *The new climate war* (Mann, 2021), argues that resistance to phasing out fossil fuels has been promoted by the fossil fuel industries, who have campaigned to persuade people and politicians that the science informing climate change is false and that switching to renewable energy will damage the economy.[12] This is the fourth layer of explanation, which could also be seen as the influence of capitalism. Students' views on which of these explanations is the most important one will determine the actions they think are needed to address climate change.

A final point to consider is that there is a danger in elevating climate change to be the dominant environmental problem of the present. There are many others that also require action, such as environmental toxins, plastic waste, land degradation and soil erosion and vegetation clearance, but which risk being ignored as attention is focused on climate.

'Geographical' environmental management strategies

Many of the strategies adopted to manage environmental problems are examples of the use of geography's key concepts and illustrate the subject's ways of thinking. For example, the creation of reserves to protect biodiversity is a spatial strategy, while the establishment of corridors to connect them is an interconnection strategy. Urban planning of the location of employment designed to reduce travel to work is another spatial strategy, while policies to increase the use of mass transit is another interconnection strategy. A major group of strategies are environmental, in that they use environmental processes, and these are discussed further. They are generally called 'nature-based solutions', which:

> involve working with and enhancing nature to help address societal challenges. They encompass a wide range of actions, such as the protection and management of natural and semi-natural ecosystems, the incorporation of green and blue infrastructure in urban areas, and the application of ecosystem-based principles to agricultural systems.
>
> (Seddon et al., 2020, p. 2)

They use the environment, rather than engineering, to manage environmental problems and can best be explained by some examples, such as:

- The restoration of natural coastal features such as mangrove forests and marshlands to protect shorelines and communities from coastal flooding and storm damage, instead of constructing seawalls (Box 6.3).
- Restoring rivers and their wetlands to manage floods, instead of constructing levees and dams. For example, in Europe, a study found that the restoration of a number of rivers reduced flood damage to crops and forests and resulted in increased agricultural production, carbon sequestration and recreation, when compared with unrestored rivers (Seddon et al., 2020, p. 3).
- Restoring vegetation along streams and rivers to reduce erosion and improve water quality.
- Protecting or restoring the ecosystems in water catchments to improve water quality, instead of constructing water treatment plants.

Box 6.3 The advantages of ecosystem-based coastal defences

Conventional coastal engineering, such as the building of sea walls, dykes and embankments, is widely perceived as the ultimate solution to combat flood risks. However, these defences are seriously challenged in many locations as their continual and costly maintenance, as well as their heightening and widening to keep up with the increasing flood risk are becoming unsustainable. Furthermore, conventional coastal engineering often exacerbates land subsidence by soil drainage and hinders the natural accumulation of sediments by tides, waves and wind, thereby compromising the natural adaptive capacity of shorelines to keep up with relative sea-level rise.

In recent years, ecosystem-based flood defence has been brought into large-scale practice as a regional solution that is more sustainable and cost-effective than conventional coastal engineering. It is applied at locations that have sufficient space between urbanized areas and the coastline to accommodate the creation of ecosystems, such as tidal marshes, mangroves, dunes, coral reefs and shellfish reefs, that have the natural capacity to reduce storm waves and storm surges, and can keep up with sea-level rise by natural accretion of mineral and biogenic sediments. The latter process secures the long-term sustainability of ecosystem-based coastal protection. Furthermore, these ecosystems provide several added benefits, including water quality improvement, fisheries production and recreation, so that in the long term they could be more cost effective than conventional defences.

(Temmerman et al., 2013, p. 79)

Conclusion

The concept of environment is a way of thinking about the environment and its significance for humans and covers a range of ideas. One is about understanding the environment as a complex web of interconnected phenomena and processes in which humans are embedded. Another is that the biophysical environment is our home, that we evolved in it and depend on it for our survival and wellbeing, and must take it seriously. However, the ways we are changing it have produced some serious environmental problems that threaten its future ability to sustain us. A third idea is whether the environment has an influence on our economies and societies. A study of the environment provides students with the opportunity to combine biophysical and human explanations of environmental problems and to appreciate how geography can support an integrated analysis of some significant contemporary issues. Furthermore, understanding the causes of these problems provides students with the knowledge to think of strategies to respond to them and to become active in the changes needed to protect humanity's future.

Summary

- The key concept of environment is about how we conceptualise and understand the physical and biological context in which we live and our interrelationships with it.
- The environment comprises the physical and biological elements of the Earth's surface, including those that people have constructed, removed or modified.
- Humans are an integral part of the environment but have a unique ability to alter it.
- A landscape is the visual appearance of an environment, and interpreting a landscape and what it reveals integrates many aspects of geography.
- The biophysical environment enables and supports human life and wellbeing in four ways.
- Environmental phenomena are connected with each other, and with people, in complex ways.
- The environment has an influence on human societies, but this is mediated by culture, technology and institutions, has weakened over time and is contested.
- Humans have been changing the environment for a very long time, but at an accelerating rate.
- Environmental problems are also social and economic problems.
- The causes of environmental problems are complex and involve several layers of explanation.
- The choice of strategies to manage an environmental problem depends on an understanding of the causes of the problem.
- Nature-based solutions use environmental processes to reduce environmental problems.
- The study of environmental phenomena and environmental change requires the integration of biophysical and human phenomena and processes.

How could you use this chapter in teaching?

Many of the ideas in this chapter could be taught through field work in, for example:

- An urban environment.
- A farm where soil is taken seriously.
- A river.
- A landscape.
- A coastal area.

Questions that you could ask your students include:

- What is the environment?
- What is the place of humans in the environment?
- How dependent are we on the environment for our wellbeing?
- Why is the key concept of interconnection so important in the study of the environment?
- To what extent, if any, has the environment influenced human societies?
- Why do we need to understand the causes of an environmental problem before choosing strategies to manage it?
- What are environmental or nature-based solutions, and why are they useful?
- Is the environment a factor in what we are studying at present?

Students could consolidate their thinking about the environment by creating statements that generalise what they have been studying, as in these examples:

> Water is a difficult resource to manage because it is integrated into environmental systems in complex ways, can be highly variable over time and across space, is essential for human, animal and biotic life and has many competing uses and values.
>
> Humans are dependent on the biophysical environment for their survival. It supports and enriches human life by providing raw materials and food, recycling and absorbing wastes, maintaining a safe habitat and being a source of enjoyment, inspiration and spirituality.
>
> People perceive, adapt to and use similar biophysical environments in different ways, because of differences in culture, population density, type of economy, level of technology and values.

Notes

1 Day's eight themes in physical geography are: 'the study of systems and processes; the study of the natural world; a spatial perspective; anthropogenic processes; impacts on people; environmental change; interconnected systems; and a scientific approach' (Day, 2017, p. 38). See also Harden et al. (2020).
2 For the UK, see Boardman (2021).
3 See Goudie (2019), Chapter 5, for a general review; and Wohl and Merritts, 2007, for the USA. Macklin and Lewin, 2019, examine human alteration of rivers from 7000BP to the present, and combine this with the effects of climatic change. They point out that even the rivers of the USA were 'non-pristine' before 1492.
4 See also Wohl et al. (2015) and the website of the European Centre for River Restoration.
5 See Nayeri (2021), and Puttick et al. (2018).
6 For a review of Marsh's contribution to physical geography and the study of human-environment relationships, see Bendix and Urban (2021).
7 For an alternative view to that of Kiage, see Ussiri and Lal (2019).
8 An example of this form of integrated physical-human research is McClintock's study of urban soil contamination in Oakland, California (McClintock, 2015).
9 For a training program on climate change, see the website of Climate Fresk, (https://climatefresk.org/), an international organisation that advertises that 'In just 3 hours, the collaborative Climate Fresk workshop will teach you the fundamental science behind climate change and empower you to take action'.
10 There is also a good review of research into the effects of climate change on wellbeing, by Adger et al. (2022).
11 The material in the last two paragraphs was first published in Maude (2022).
12 For an explanation of the reasons people and businesses reject or criticise the arguments for climate change, see Hornsey and Lewandowsky (2022).

Useful reading

Evans, M. (2021). *Soil: The incredible story of what keeps the earth, and us, healthy.* Murdoch Books.
A very readable and entertaining explanation of the value of soil.

Gergis, J. (2022). *Humanity's moment: A climate scientist's case for hope.* Black.
A book by a climate scientist whose first degree was in physical geography that explains the science and effects of global warming, what needs to be done to prevent temperatures from rising beyond a safe level, and what it is like to be a climate scientist.

Goudie, A. S. (2019). *Human impact on the natural environment: past, present and future* (8th ed.). John Wiley.
A comprehensive book by a leading British physical geographer.

Harden, C. P., Luzzadder-Beach, S., MacDonald, G. M., Marston, R. A., & Winkler, J. A. (2020). Physical geography contributes. *Progress in Physical Geography, 44*(1), 5–13. https://doi.org/10.1177/0309133319893918
Discusses the relationships between physical geography and some of the concepts in this book.

Holden, J. (2011). *Physical geography: The basics.* Routledge.
A short and clear introduction to physical geography for those who know they need it. The summaries at the end of each chapter are excellent examples of generalisations.

Larsen, T. B., & Harrington, J. (2020). *Enhancing powerful geography with human-environment geography.* Texas State University. https://digital.library.txstate.edu/handle/10877/16158

Middleton, N. (2012). *Rivers: A very short introduction.* Oxford University Press.
A wide-ranging and readable book on rivers by a geographer.

Pedersen Zari, M., MacKinnon, M., Varshney, K., & Bakshi, N. (2022). Regenerative living cities and the urban climate–biodiversity–wellbeing nexus. *Nature Climate Change, 12*, 596–606. https://doi.org/10.1038/s41558-022-01390-w
A short review of nature-based solutions for urban environments.

References

Adger, W. N., Barnett, J., Heath, S., & Jarillo, S. (2022). Climate change affects multiple dimensions of well-being through impacts, information and policy responses. *Nature Human Behaviour, 6*, 1465–1473. https://doi.org/10.1038/s41562-022-01467-8

Aldrian, E (2021). Indonesia's capital Jakarta is sinking. Here's how to stop this. *The Conversation*, November 11. https://theconversation.com/indonesias-capital-jakarta-is-sinking-heres-how-to-stop-this-170269

Amundson, R., Berhem A. A., Hopmans, J. W., Olson, C., Sztein, A. E., & Sparks, D. L. (2015). Soil and human security in the 21st century. *Science, 348*, 1261071. https://doi.org/10.1126/science.1261071

Ashmore, P. (2015). Towards a sociogeomorphology of rivers. *Geomorphology, 251*, 149–156. https://doi.org/10.1016/j.geomorph.2015.02.020

Behnke, R., & Mortimore, M. (2016). Introduction: The end of desertification? In R. H. Behnke & M. Mortimore, *The end of desertification? Disputing environmental change in the drylands* (pp.1–33). Springer Verlag.

Bendix, J., & Urban, M. A. (2021). Nothing new under the sun? George Perkins Marsh and roots of U.S. physical geography. *Annals of the American Association of Geographers, 111*(3), 709–716. https://doi.org/10.1080/24694452.2020.1761769

Boardman, J. (2021). How much is soil erosion costing us? *Geography, 106*(1), 32–38. https://doi.org/10.1080/00167487.2020.1862584

Bounoua, L., Zhang, P., Nigro, J., Lachir, A., & Thome, K. (2017). Regional impacts of urbanization on the United States. *Canadian Journal of Remote Sensing, 43*, 256–268. https://doi.org/10.1080/07038992.2017.1317208

Brierley, G., Tadaki, M., Hikuroa, D., Blue, B., Šunde, C., Tunnicliffe, J., & Salmond, A. (2019). A geomorphic perspective on the rights of the river in Aotearoa New Zealand. *River Research Applications, 35*, 1640–1651. https://doi.org/10.1002/rra.3343

Bürgi, M., Hersperger, A. M., & Schneeberger, N. (2004). Driving forces of landscape change – Current and new directions. *Landscape Ecology, 19*, 857–868. https://doi.org/10.1007/s10980-005-0245-3

Davidson, N. C. (2014). How much wetland has the world lost? Long-term and recent trends in global wetland area. *Marine and Freshwater Research, 65*, 934–941. https://www.publish.csiro.au/MF/MF14173

Day, T. (2017). Core themes in textbook definitions of physical geography. *Canadian Geographer, 61*(1), 28–40. https://doi.org/10.1111/cag.12354

Gallup, J. L. & Sachs, J. D. with Mellinger, A. D. (1998). Geography and economic development. *Annual World Bank Conference on Development Economics*, 127–78.
Geertz, C. (1963). *Agricultural involution: The process of ecological change in Indonesia*. University of California Press.
Ghazoul, J. (2015). *Forests: A very short introduction*. Oxford University Press.
Gibbs, L. M. (2010). "A beautiful soaking rain": Environmental value and water beyond Eurocentrism. *Environment and Planning D: Society and Space*, *28*, 363–378. https://doi.org/10.1068/d9207
Gilvear, D. J., & Jeffries, R. (2017). Fluvial geomorphology and river management. In J. Holden (Ed.), *Physical geography and the environment* (4th ed., pp. 493–524). Pearson Education.
Goudie, A. S. (2019). *Human impact on the natural environment: Past, present and future* (8th ed.). John Wiley.
Harden, C. P. (2014). The human-landscape system: Challenges for geomorphologists. *Physical Geography*, *35*(1), 76–89. https://doi.org/10.1080/02723646.2013.864916
Hornsey, M. J., & Lewandowsky, S. (2022). A toolkit for understanding and addressing climate scepticism. *Nature Human Behaviour*, *6*, 1454–1464. https://doi.org/10.1038/s41562-022-01463-y
Huggett, R., & Perkins, C. (2007) Place as landscape. In I. Douglas, R. Huggett, & C. Perkins (Eds.), *Companion encyclopedia of geography: From local to global, vol. 1* (2nd ed., pp. 17–30). Routledge.
Jones, S. (2008). Political ecology and land degradation: How does the land lie 21 years after Blaikie and Brookfield's land degradation and society? *Geography Compass*, *2*(3), 671–694. https://doi.org/10.1111/j.1749-8198.2008.00109.x
Kanemoto, K., Moran, D., Lenzen, M., & Geschke, A. (2014). International trade undermines national emission reduction targets: New evidence from air pollution. *Global Environmental Change*, *24*, 52–59. https://doi.org/10.1016/j.gloenvcha.2013.09.008
Kiage, L. M. (2013). Perspectives on the assumed causes of land degradation in the rangelands of Sub-Saharan Africa. *Progress in Physical Geography*, *37*(5), 664–684. https://doi.org/10.1177/0309133313492543
Lane, S. N. (2019). Critical physical geography. *Geography*, *104*(1), 49–53. https://doi.org/10.1080/00167487.2019.12094062
Larsen, T., Gerike, M., & Harrington, J. (2021). Human-environment thinking and K-12 geography education. *Journal of Geography*. https://doi.org/10.1080/00221341.2021.2005666
Legates, D. R., Mahmood, R., Levia, D. F., DeLiberty, T. L., Quiring, S. M., Houser, C., & Nelson, F. E. (2011). Soil moisture: A central and unifying theme in physical geography. *Progress in Physical Geography*, *35*, 65–86. https://doi.org/10.1177/0309133310386514
Lewin, J., & O'Shea, T. (2023). Floodplains: Fusing analysis of the territorially constructed with analysis of natural terrain processes. *Progress in Physical Geography* (online). https://doi.org/10.1177/03091333231156510
MacKerron, G., & Mourato, S. (2013). Happiness in greater in natural environments. *Global Environmental Change*, *23*, 992–1000. https://doi.org/10.1016/j.gloenvcha.2013.03.010
Macklin, M. G., & Lewin, J. (2019). River stresses in anthropogenic times: Large-scale global patterns and extended environmental timelines. *Progress in Physical Geography*, *49*(1), 3–23. https://doi.org/10.1177/0309133318803013

Mann, M. E. (2021). *The new climate war: The fight to take back our planet.* Scribe.
Marsh, G. P. (2003). *Man and nature, or, physical geography as modified by human action.* University of Washington Press.
Maude, A. (2022). Using geography's conceptual ways of thinking to teach about sustainable development. *International Research in Geographical and Environmental Education.* https://doi.org/10.1080/10382046.2022.2079407
McClintock, N. (2015). A critical physical geography of urban soil contamination. *Geoforum, 65,* 69–85. https://doi.org/10.1016/j.geoforum.2015.07.010
Meyer, W. B., & Guss, D. M. T. (2017). *Neo-environmental determinism: Geographical critiques.* Springer Nature.
Montanarella, L., Scholes, R., & Brainich, A. (Eds.) (2018). *The IPBES assessment report on land degradation and restoration.* Secretariat of the Intergovernmental Science-Policy Platform on Biodiversity and Ecosystem Services.
Nayeri, C. (2021). Unnatural hazards: Multiplying the questions we ask. *Geography, 106*(2), 85–91. https://doi.org/10.1080/00167487.2021.1919411
Olsson, L. Barbosa, H., Bhadwal, S., Cowie, A., Delusca, K., Flores-Renteria, D., Hermans, K., Jobbagy, E., Kurz, W., Li, D., Sonwa, D. J., & Stringer, L. (2019). Land Degradation. In P. R. Shukla, J. Skea, E. Calvo Buendia, V. Masson-Delmotte, H.-O. Pörtner, D. C. Roberts, P. Zhai, R. Slade, S. Connors, R. van Diemen, M. Ferrat, E. Haughey, S. Luz, S. Neogi, M. Pathak, J. Petzold, J. Portugal Pereira, P. Vyas, E. Huntley … J. Malley (Eds.), *Climate change and land: An IPCC special report on climate change, desertification, land degradation, sustainable land management, food security, and greenhouse gas fluxes in terrestrial ecosystems* (pp. 345–436). IPCC.
Pielke, R. A., Mahmood, R., & McAlpine, C. (2016). Land's complex role in climate change. *Physics Today, 69*(11), 40–46. https://doi.org/10.1063/PT.3.3364
Puttick, S., Bosher, L., & Chmutina, K. (2018). Disasters are not natural. *Teaching Geography, 43*(3), 118–120. https://portal.geography.org.uk/journal/view/J003094
Rojas-Rueda, D., & Nieuwenhuijsen, M. J. (2019). Green spaces and mortality: A systematic review and meta-analysis of cohort studies. *Lancet Planetary Health, 3,* 469–477. https://doi.org/10.1016/S2542-5196(21)00229-1
Seddon, N., Chausson, A., Berry, P., Girardin, C. A. J., Smith, A., & Turner, B. (2020). Understanding the value and limits of nature-based solutions to climate change and other global challenges. *Philosophical Transactions of the Royal Society B, 375,* 20190120. https://doi.org/10.1098/rstb.2019.0120
Selmes, I. (2017). Sensual soils. *Teaching Geography, 42*(3), 93–95. https://portal.geography.org.uk/journal/view/J002956
Sheppard, E. (2022). Geography and the present conjuncture. *EPF: Philosophy, Theory, Models, Methods and Practice, 1*(1) 14–25. https://doi.org/10.1177/26349825221082164
Temmerman, S., Meire, P., Bouma, T. J., Herman, P. M. J., Ysebaert, T., & De Vriend, H. J. (2013). Ecosystem-based coastal defence in the face of global change. *Nature, 504*(7478), 79–83. https://doi.org/10.1038/nature12859
Thomas, W. L. (Ed.) (1956). *Man's role in changing the face of the earth.* University of Chicago Press.
Tol, R. S. J. (2021). The economic impact of climate in the long run. In A. Markandya & D. Rübbelke (Eds.), *Climate and development* (pp. 3–36). World Scientific. https://doi.org/10.1142/9789811240553_0001

Urban, M., & Rhoads, B. (2003). Conceptions of nature: Implications for an integrated geography. In S. Trudgill & A. Roy (Eds.), *Contemporary meanings in physical geography: From what to why?* (pp. 211–231). Hodder Education.
Urban, M. A. (2018). In defense of crappy landscapes (core tenet #1). In R. Lave, C. Biermann, & S. N. Lane (Eds.), *The Palgrave handbook of critical physical geography* (pp. 49–66). Palgrave Macmillan.
Ussiri, D. A. N., & Lal, R. (2019). Soil degradation in Sub-Saharan Africa: Challenges and opportunities. In R. Lal & B. A. Stewart (Eds.), *Soil degradation and restoration in Africa* (pp. 1–24). Taylor & Francis Group.
Wohl, E., Lane, S. N., & Wilcox, A. C. (2015). The science and practice of river restoration. *Water Resources Research*, *51*, 5974–5997. https://doi.org/10.1002/2014WR016874
Wohl, E., & Merritts, D. J. (2007). What is a natural river? *Geography Compass*, *1*(4), 871–900. https://doi.org/10.1111/j.1749-8198.2007.00049.x
Wulf, A. (2015). *The invention of nature: The adventures of Alexander Von Humboldt, the lost hero of science.* John Murray.
Wylie, J. (2011). Landscape. In J. A. Agnew & D. N. Livingstone (Eds.), *The SAGE handbook of geographical knowledge* (pp. 300–315). SAGE Publications.

Chapter 7

The analytical concepts

Scale and time

In Chapter 2 scale and time were described as analytical concepts, because their main functions are to organise and analyse information, and they cannot be used to directly explain. This chapter discusses some of the ways that geographers conceptualise and use scale and time.

Conceptualising scale

The academic geographical literature on scale can be complex, confusing and contradictory, and much of it is not applicable to school geography. This section of the chapter goes beyond the usual explanations of scale in many school textbooks into some of this literature, but selectively.

Geography in schools uses two types of scale—cartographic and spatial. *Cartographic scale* refers to the numerical relationship between distance on a map and the corresponding distance on the surface of the Earth. This relationship is usually expressed as a representative fraction, such as 1:10,000, which means that 1 unit of distance on the map corresponds to 10,000 units of distance on the Earth. Cartographic scale can also be represented in words, such '1 cm equals 10 kms', or graphically by a line on the map that is equivalent to a specified distance on the ground. The graphical method has the advantage that if a map is enlarged or reduced the real distance the line represents remains correct, whereas the representative fraction will no longer be correct because distance on the map has changed, but not distance on the Earth.

A map with a scale of 1:500,000 covers a relatively large area of land compared to a map with a scale of 1:10,000. This means that maps with a small scale (i.e., a small fraction) portray larger areas than maps with larger scales (i.e., a larger fraction), which portray smaller areas. This is a cartographic convention that can be a source of confusion, as it is reversed for *spatial scale* which refers to the areal size selected to observe the world. In geography this generally varies from a small area such as the local (the reach of a river, a neighbourhood, a small town or a landscape), to the regional (such as a river catchment, large city or region), national and global. A study of a small area is then called a small-scale study, while a large-scale study might refer to a nation or a major river basin.

DOI: 10.4324/9781003376668-8

In physical geography these areal units are defined by biological and physical features, as interpreted by scientists. In human geography they are better understood as human constructs about how to classify the ways that space can be subdivided. For example, the nation state is a relatively recent creation, and did not exist for most of human history. Similarly, the growing focus on the global scale and globalisation processes is also relatively recent, and a product of the technological innovations that have enabled a growing number of businesses to operate globally. The term 'globalisation' was first used in 1930, and describes the increasing economic integration of countries, and the rise of supranational corporations and financial institutions that were thought to be beyond the control of national governments. While trade between world regions goes back at least to the First Century BCE, with the Silk Road linking China with Europe, the term globalisation is much more recent. The nation and the global are consequently scales that have been produced by the actions of people, businesses and governments. Herod describes this view of scales as:

> scales do not just exist, waiting to be utilized, but must instead be brought into being. Thus, transnational corporations do not simply adapt their activities to a pre-made global scale defined by the Earth's geologically given limits but must, instead, actively build their own global scale of operation. They must, in effect, become 'global'.
>
> (Herod, 2009, p. 219)

The choice of scale

Geographers often describe the use of scale as like looking through a zoom lens, with which one can focus on a very small area, or zoom out to a much larger view. What is significant about this is that what one observes, or finds out, changes as one zooms in and out. Life expectancy in the USA (Table 7.1) illustrates this. At a national level life expectancy was 78.8 years in 2019, and the USA ranked about 40th in the world, well behind many less wealthy countries. At the level of the states the lowest life expectancy was 74.8, roughly similar to that in Belarus. At the level of counties, the lowest life expectancy was 66.8 years, roughly equivalent to that in Uganda. Finally, at the level of census tracts, with populations generally between 1200 and 8000, the lowest life expectancy was 56.9 years, which is roughly equivalent to that in Somalia, a country with the third lowest life expectancy in the world. The difference between the areas with the highest and lowest life expectancy increased from 7.5 years at the state scale to 20.0 years at the scale of counties and 40.6 years at the scale of census tracts. What the table reveals is that observations at a national or state level conceal considerable variation at smaller scales, a conclusion that applies to a very wide range of socioeconomic data. In the academic literature this is called the Modifiable Areal Unit Problem, which describes situations in which findings differ according to the scale of analysis.

Table 7.1 Life expectancy in the USA

Jurisdiction	*Life expectancy at birth in years*
National, 2019	
USA	78.8
States, 2019	
Hawaii	82.3
West Virginia	74.8
Counties, 2014	
Summit County, Colorado	86.8
Ogalala Lakota County, South Dakota	66.8
Census Tracts, 2010–2015	
201.04, Chatham County, North Carolina	97.5
9569, Logan County, West Virginia	56.9

Sources: National Centre for Health Statistics, Centers for Disease Control and Prevention (https://www.cdc.gov/nchs/data-visualization/life-expectancy/index.html); Institute for Health Metrics and Evaluation (https://www.healthdata.org/us-health/data-download).

This applied not only to single variables such as life expectancy but also to the relationship between two variables, which can be different at different scales.

Vegetation types provide an environmental example of the same idea. At a global scale the world's vegetation can be divided into a number of biomes, such as northern coniferous forest, temperate grassland, tropical rainforest and desert. At this scale the whole of the UK is classified as temperate deciduous forest. At the scale of the nation, however, five major ecosystems can be identified in the UK (woodlands, moorlands, heathlands, lowland raised bogs and lowland fen), only one of which corresponds to temperate deciduous forest. At much smaller scales an even finer subdivision of vegetation can be distinguished, and an example of this is the vertical zonation of vegetation types on Mount Lemmon in Arizona described in Chapter 3.

The choice of scale is also important in studies that are investigative rather than descriptive and are about understanding, or finding answers, because the scale selected in these studies must match the problem being investigated. Gentrification, for example, should be studied at the scale of that part of an urban area in which the process occurs. Studies of flooding require investigation at the scale of the catchment, which is bigger than the area flooded but is the area producing the water for a flood, while studies of landslides are at the scale of individual events. What this means is that the scale of a study may be determined by the phenomenon being studied.

> Rather than scale being absolute within a fixed reference frame with the entities fitting within this, scale could be thought of as being defined by the entities themselves. … The entities themselves define the processes or flows

> forming them, they define the spatial and temporal dimensions of importance rather than being defined by these.
>
> (Inkpen and Wilson, 2005, p. 163)

This comment is about physical geography, but it can also apply to human geography, where it illustrates the contention that scale does not exist independently of human actions and the processes they create, such as gentrification.

The choice of scale not only influences the patterns that will be discovered but also the explanations for these patterns. 'It has long been known that generalizations made at one level do not necessarily hold at another, and that conclusions derived at one scale may be invalid at another' (Burt, 2009, p. 201). For example, natural vegetation at the global scale is largely controlled by climate, but at smaller scales vegetation patterns are determined by factors such as topography, aspect, soil and drainage. In the case of climate, at the global scale climates are classified into the familiar types identified by Köppen and his collaborators, and are based on temperature and precipitation, but at smaller scales, they are modified by altitude, topography (depressions collect cooler air moving down slopes), vegetation, aspect (whether facing the sun or not), nearness to the sea, and urban development. Similarly, studies show that the main influence on soil erosion at the very smallest scale is soil structure, but at the catchment scale it is land use and vegetation.

Tim Burt also describes a more general problem in the use of scale in research when he writes that 'Analysis encourages the use of the microscope while synthesis requires the wide-angle lens!' (Burt, 2009, pp. 212–213). Two examples relevant to school geography may help to illustrate this problem. Understanding the causes of deforestation in the Amazon Basin requires small-scale field work with the people and industries that are clearing the forest, in order to understand the factors (which will range from local to global) that are causing their actions. These will vary throughout the Amazon and over time and require a number of small-scale studies to identify. This research can identify the mechanisms that cause deforestation, but then must be related to the wider picture of trends in deforestation within the whole of the Amazon Basin, obtained from satellite images, to understand the whole. A second example is that of migration. Research at the scale of regions and sub-regions might find a relationship between poverty and out-migration, with migration more likely from poorer than richer places. However, research at the individual or household level may find that the poorest people are less likely to migrate than middle-income people, because they lack the resources and skills, and if they do migrate, it is to closer destinations.

Scalar 'traps'

There are several other issues to be aware of when working with scales, as these can affect one's perceptions and judgements about the direction of influences and the location of actions. For example, in the conventional hierarchy of scales from the neighbourhood to the world the global scale may be perceived as the dominant one, controlling or influencing those below, while the local scale may be perceived as weak and powerless. This perception is frequently used by neoliberal governments to argue that globalisation is inevitable and irresistible, and consequently that policies to protect local industries and places from the effects of globalisation are pointless. Critics of this viewpoint, on the other hand, argue that it is possible for local people to challenge and resist global actions, for example by preventing the establishment of a McDonald's outlet or a mine in their locality. The perception of the local as lacking power also conflicts with geographers' views on the importance of place, and the use of place to ground environmental, economic and social programs, as discussed in Chapter 4. Critics might also point out that the global economy is simply the cumulative result of decisions made by people who live in local places. Furthermore, they might argue that global firms have to locate their operations in local places and that:

> manufacturers may have to 'localize' themselves by developing business linkages with local suppliers, by training a community's workforce to operate particular types of machinery used in their plant, by establishing credit with a community's banks and other financial institutions, and by building trust with politicians who represent the community in different levels of government.
>
> (Herod, 2009, p. 220)

Should the global then be viewed as multi-local?

Another 'trap' is that people may perceive global issues, such as climate change, as requiring action at the global level, and consequently neglect solutions at the local scale because they are thought to be too small to have an impact on the problem. Others will respond that global problems have to be addressed at all scales, but in different ways, and that local actions are an essential part of any solution. They might also point out that 'negotiations at the global level often stall and achieve little impact because of the impossibility to satisfy the interests and concerns of all stakeholders' (Dittrich, 2022, p. 189). As Thomas Wilbanks and Robert Kates write:

> In theory, scale matters in studying global change, local dynamics are worth worrying about, and localities can make a difference. For instance, it is clear that some of the driving forces for global change operate at a global scale, such as the greenhouse gas composition of the atmosphere and the reach of global financial systems. But it seems just as clear that many of the individual phenomena that underlie microenvironmental processes, economic activities, resource use, and population dynamics arise at a local scale.
>
> (Wilbanks and Kates, 1999, p. 602)

A different issue to that of the domination of the global is one that has been described as the 'local trap', in which researchers 'assume that the key to environmental sustainability, social justice, and democracy ... is devolution of power to local-scale actors and organizations'. This assumption is based on a belief that when decisions are made by local people 'the outcome will be more socially and environmentally just than if the decisions are made, for example, by a national body politic' (Brown and Purcell, 2005, p. 608). This assumption is not necessarily correct.

A hierarchical view of scale may also lead to an assumption that decisions flow down from the global to the national and then to local scales. What such a view misses is the growing number of states, regions, cities and other entities that have formed independent networks across the same scale. The most significant are perhaps the networks of states and cities focused on action for climate change, often formed because of frustration with the lack of action by national governments, as in the USA and Australia. Some examples are:

- The US Climate Alliance. Formed in 2017 in response to President Trump's withdrawal of the USA from the Paris Agreement, it is a bipartisan coalition of 25 state governors working to achieve the goals of the Paris Agreement.
- The Under2 Coalition. Formed in 2015, it is a network of 260 state and regional governments across the globe representing 50% of the world's economy, also working to reduce emissions in line with the Paris Agreement.
- Climate Mayors, founded in 2014, is a bipartisan network of more than 470 US city mayors committed to climate leadership through actions in their communities.

These are examples of states and cities networking with each other and sometimes in opposition to the policies of their national governments. Their role, and the implications for thinking about scale, is concisely expressed by the authors of a study of the climate change policies of two US cities, Portland and Tulsa, in which they claim that:

> Cities are not only formulating their own policies on greenhouse gas emissions. They also participate in national and international networks of cities engaging the problem of climate change and interact on a day-to-day basis with the multinational corporations which they host. This simultaneous interface of the local, national, and global raises questions of cities' scale in a globalizing world.
>
> (Osofsky and Levit, 2008, p. 413)

This passage draws attention to the ways that networks not only link entities at the same scale but also engage with entities at other scales. Such networks are not confined to cities and states, but may also be formed by environmental groups, trade unions, Indigenous peoples and others who need to engage with organisations and governments at higher scales to achieve their aims. Scott Hoefle describes the example of the Coordenação das Organizações Indígenas da Amazônia Brasileira, a consortium of Amerindian social movements in the Brazilian Amazon formed in 1989. It is a multi-scalar organisation combining bottom-up political participation with alliances with government and non-government organisations at all scales.

> [It] has been highly successful at setting aside enormous areas of the Amazon for reservations and at attracting funding for health and community development programmes. Representatives are regularly called upon to testify before state and federal congressional committees, to participate in ministry commissions and to attend all of the important world environmental and nongovernmental association events.
>
> (Hoefle, 2006, p. 240)

Another concern is that studies at one scale may miss explanations that derive from other scales. This problem was noted in Chapter 2, with the idea of levels of explanation. For example, the low average income of a place could be the result of a lack of local resources (local scale), the economic policies of the national government (national scale), or the decisions of a distant corporation to close a local factory (global scale).

A similar problem is that the effects of some action or policy may only be perceived at one scale and the effects at other scales missed or ignored. An example is the fracking of shale to produce natural gas. In the early years of the industry, fracking was supported by some environmental groups because natural gas produces less carbon dioxide per unit of energy than coal, and they were thinking only at a national or global scale about climate change. It took time for the environmental and health effects of fracking to become evident at the local scale and for the public debate to shift towards greater controls on the industry.

The ecological fallacy is a final trap. The trap is that relationships between groups identified from aggregated data cannot be used to make inferences about the individuals who have been aggregated. This is because aggregated data are averages, which conceal the variations between individuals. An example is the relationship between poverty and migration noted earlier. Research might find a relationship between the two at the scale of local areas, but not at the scale of individual people or households.

An understanding of scale is of fundamental importance in school geography, but teachers need to be aware of some of its complexities and of the assumptions about scale that may limit students' thinking.

Time

Time has been classified as an analytical concept because it can be used as a framework for understanding the present and thinking about the future. Time itself cannot explain, but examining how phenomena have developed over time can help in understanding their present characteristics or forecasting their future ones. Monitoring the state of the environment also requires an investigation of change over time. However, perhaps the most valuable aspect of incorporating time into school geography is that it gives young people an awareness of change and an understanding that the future is unlikely to be the same as the present. Charles Rawding argues that:

> notions of time must be central to the teaching of Geography and ... geographical outcomes have to be seen as a consequence of historical processes. The need to emphasise time is reinforced when one considers the limited experience of school pupils. Even a 16-year-old pupil is unlikely to have geographically useful recollection of events that are more than six or seven years old. Put simply, schoolchildren have not lived long enough to realise that change is the norm and that understanding change over time is a fundamental element in geographical understanding.
>
> (Rawding, 2013, p. 45)

Students should also be aware of the concept of time that they will learn from their society. In Western cultures time is linear and associated with the idea of 'progress', and each new year is celebrated as a new beginning. In many Indigenous societies, on the other hand, time is seasonal and circular, and there is no progression from one year to the next and no expectation that next year will be different. This is a very different view of time. Our sense of time has also changed over the past centuries. Agricultural societies worked by the sun, as did our hunter-gatherer ancestors for most of human history, but industrial societies needed the clock to coordinate work, school and transportation, and time became precisely measured and standardised. It has also speeded up, as the time taken to travel distances and obtain information has decreased. Trains, steamships and then aircraft have reduced the time taken to cover distance, while telecommunications technologies now make information quickly accessible, producing expectations that messages will be responded to instantly. This trend has been called time-space convergence, and was explained in Chapter 5.

Timescales

There are several timescales that students of geography should know something about, because they will help them to understand processes of change that contribute to the present. The three discussed here are the Holocene, climate change and technological change since the Industrial Revolution.

The Holocene

The geological epoch called the Holocene began around 12,000 years ago, and is the current epoch until there is a formal agreement to replace it with the Anthropocene. The Holocene has been a period of warmer and relatively stable climate following the end of the last glaciation, although temperatures declined slightly between around 1350 and 1900, a period known as the Little Ice Age. Ian Lawson describes some of the changes this produced in Britain:

> Rivers and lakes in Britain ... froze much more regularly and thickly than they do today; people held fairs on the ice on the River Thames in London, the farmers of East Anglia became famous for their prowess in ice-skating on their frozen ditches, and the Scots developed the sport of curling, played on the ice of frozen lakes.
>
> (Lawson, 2017, p. 117)

The Holocene has been a period of significant environmental change and societal development. These changes include:

Rising sea levels

As the ice melted sea levels rose around the world, drowning coastal areas and creating most of the world's present coastal geomorphological features. Around Australia the rise was about 120 m from the Last Glacial maximum around 20,000 years ago, and this submerged 23% of the then Sahul land mass, consisting of what is now Australia and the island of New Guinea (Nunn, 2018, pp. 244–245). Patrick Nunn records that Aboriginal communities around the present coast of Australia still have stories about inundation by rising sea levels that can be dated from geomorphological evidence to events some 7000 to 13,000 years ago. Similarly, Lawson (2017, p. 119) writes that some paleogeographers believe that the Black Sea was once a smaller lake, which flooded catastrophically when seal level rise in the Mediterranean overtopped the sill at what is now the Bosporus Strait, 'an event so cataclysmic for local populations that some people suggest that it inspired myths and legends of catastrophic floods in many Near Eastern cultures, persisting to this day in the Biblical story of Noah'.

The outcomes of sea level rise include drowned river valleys, some of which have become major harbours, islands that were once hills and mountain ranges, and sedimentation of the lower reaches of rivers resulting from the decline in the gradients of river systems as their base level rose.

Isostatic rebound

When the ice sheets melted their weight was removed from the land, which then began to rise slowly. Where uplift was faster than sea level rise the result has been raised beaches. Some parts of the world are still rising, as in Scotland.

Soil development

As temperatures increased soils developed through physical and chemical weathering and the biological effects of the growth of vegetation.

Agriculture and population growth

The beginning of the Holocene coincided with the development of agriculture (and may have made it possible) and a gradual but accelerating impact of humans on the environment.

Land cover change

Land cover refers to the cover of the surface of the Earth, such as ice, water, natural vegetation, cropland and urban areas. Pre-Holocene societies of hunter-gatherers made some changes to land cover, mainly by burning vegetation to manage game. From about 1700 CE onwards, however, population growth, economic development and the commercialisation of agriculture and forestry resulted in the expansion of cropland from 3% to 15% of the habitable land surface of the globe by 2018, and of pasture from 6% to 31%, according to one study. At the same time the area under forest and woodland shrank from 52% to 38%. These changes to land cover not only had impacts on local environments, such as a decline in rainfall and an increase in soil erosion, but are now also contributing to global warming. These issues will be discussed in Chapter 10.

Climate change

The cooling of the Little Ice Age was reversed by the warming effects of the burning of fossil fuels, with temperatures starting to rise again in the early 1900s. In Figure 7.1, one line in the graph plots observed annual average surface temperatures from 1850 to 2020, measured as relative to 1850–1900, and shows that these had increased by over 1°C by 2020. The graph also shows the modelling used by the Intergovernmental Panel on Climate Change (IPCC) to test whether this warming trend has been most likely caused by human activities, principally the burning of fossil fuels. The 'simulated natural only' line shows the estimated temperatures that would have resulted from variations in solar and volcanic activity, while the 'simulated human and natural' line adds the effects of human activities and demonstrates that the warming trend cannot be explained by natural causes alone. Note that the two simulations have been extended back to 1850, to validate them by seeing whether they are able to accurately predict the past. This graph is the most compelling evidence for the effects of fossil fuels on global warming, and the modelling on which it is based can be used to forecast future temperatures based on assumptions about future greenhouse gas emissions.

Scientists warn that the Holocene was a time when climates were favourable for human flourishing, and that 'A continuing trajectory away from the Holocene could lead, with an uncomfortably high probability, to a very different state of the Earth system, one that is likely to be much less hospitable to the development of human societies' (Steffen et al., 2015, 1259855-1). This is an important idea that students should understand; that humanity appears to be

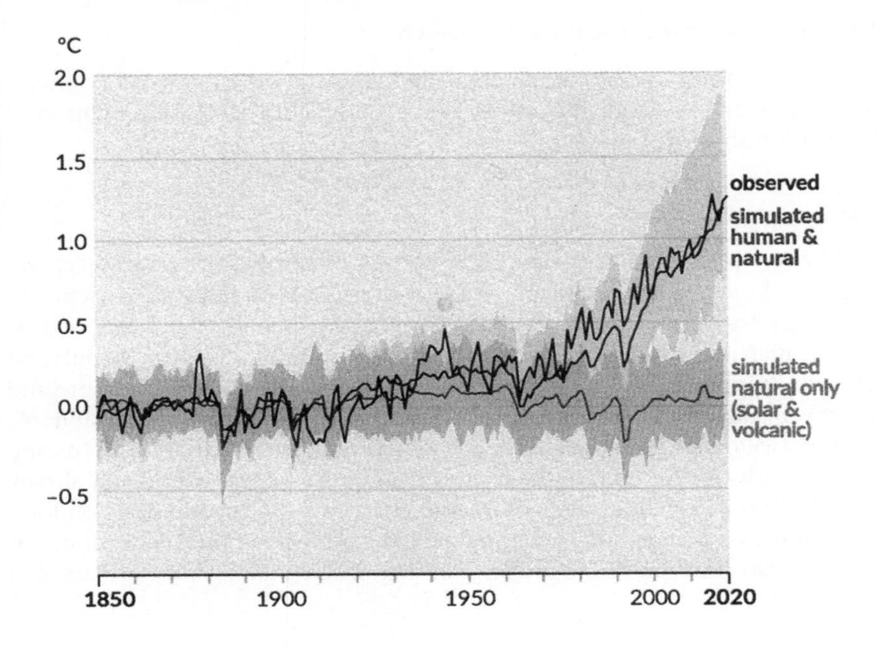

Figure 7.1 Change in global surface temperature (annual average) as observed and simulated using human & natural and only natural factors (1850–2020).

Source: Figure SPM. 1(b) from IPCC (2021, p. 6).

moving from a climate that was relatively benign and stable to one that will be less favourable for human wellbeing.

Technological change

Since about the 1780s, the economic history of the world has been marked by successive technological revolutions that have disrupted and reshaped the geography of more and more countries. Part of this pattern was first identified in the 1920s by a Russian economist, Nicolai Kondratieff (1892–1938),[1] who argued that economic growth in Europe between the 1780s and the 1920s had occurred in three long waves or cycles of prosperity, separated by periods of economic depression. He also noted that between each wave 'an especially large number of important discoveries and inventions in the technique of production and communication are made, which, however, are usually applied on a large scale at the beginning of the next long upswing' (Kondratieff, 1979, p. 536). Later writers have added to his ideas, with additional waves and more emphasis on the association of each wave with particular technological innovations. A contemporary interpretation is shown in Table 7.2.

Table 7.2 Kondratieff cycles

Kondratieff cycles	*1st*	*2nd*	*3rd*	*4th*	*5th*
Period	*1780s–1830s*	*1830s–1890s*	*1890s–1930s*	*1930s–1970s*	*1970s–2000?*
Invention	Water power, iron	Steam power, steel, steamships, railway engines	Electric power, chemicals, early internal combustion engines	Automobiles, petrochemicals, aircraft, synthetic materials	Electronics, information technology, telecommunications, biotechnology
Location (selective)	UK (Lancashire, Black Country), USA (New England)	UK (South Wales, NE England, Central Scotland), Germany, (Ruhr), France, Belgium, USA	UK (West Midlands, Greater London), Germany (Hessen), France, Belgium, USA (Midwest)	UK (West Midlands, Greater London), USA, Germany, France, Sweden, Japan, South Korea	UK (Cambridge, M4 corridor), US (California), Japan, China, Taiwan, India (Bangalore)

Sources: Dicken (2011, p. 79); Pike et al. (2006, p. 83).

What is geographically interesting about these waves of technological innovation is that their locations changed with each wave, both between countries and within countries. Internationally, the shift was from the UK to Europe and the USA, and later to North-East Asia (Japan, South Korea, Taiwan and China). Within the UK industrial growth shifted from Lancashire, the Midlands and North-East England in the first and second cycles to the South of England by the fifth, while in the USA, the shift was from New England in the first cycle to California by the fifth, with the Middle Atlantic states and the Midwest in between. Some of the places that had grown during one Kondratieff wave found it difficult to adapt to the industries of the next, because they lacked the skilled workers, the knowledge and the culture required for change. This constraint has been described as 'path dependence', and it can be illustrated by the experience of old industrial regions such as the Ruhr Valley in Germany and the North-East of England. Danny Mackinnon and Andrew Cumbers explain that:

> The industrial cultures of these areas seemed to have become rigid and fossilized, meaning that firms and institutions were tied to obsolete production systems and methods, militating against more positive responses to economic change in terms of generating new products and methods.
> (Mackinnon and Cumbers, 2007, p. 39)

For students, an awareness of these changes will help them to understand the changing fortunes of cities and regions in many countries. It will also help them to understand some of the strategies being followed by many countries to develop centres of scientific and technology innovation in selected cities, often through the creation of science parks.

The fifth cycle is thought to have ended at the beginning of the 2000s, and students could try to identify the innovations of the sixth. Ones that are becoming apparent include pharmaceutical and medical innovations, nanotechnology, biotechnology, green technology, non-fossil fuels, artificial intelligence and robotics. Will the development of these technologies be dominated by Europe, North America and North-East Asia, or will the geography of innovation change once again? Measured by the number of patents and scientific articles published in recent years, the top cities for innovation are Tokyo, New York City, Seoul, Osaka, Paris, Los Angeles, London, Beijing and Shanghai. They are followed by cities in Russia, China, India, Latin America and the Middle East (West Asia), suggesting some further spatial spread of leadership in technological innovation. Innovations, however, don't always create industries in the same locations. Australia is quite good at inventing new technologies and also at selling them to another country to manufacture.

The three timescales described above will provide students with an awareness of some significant periods in the world's environmental and human history and of change as a constant element in our lives. Students should also

Table 7.3 Urban population concentration, selected countries, 2021

	Australia	*USA*	*Canada*	*UK*	*Germany*
Percentage of total population in urban areas	86	83	82	84	77
Percentage of total population in urban agglomerations of more than one million	61	47	46	27	10

Source: World Bank.

recognise that the past can constrain future change, as in the concept of path dependence mentioned above. Another example of this concept is that cities have infrastructure, land subdivisions and buildings which constrain how rapidly urban places can change, or which make old industrial areas unattractive to new industries.

Using the past to understand the present

The example described in this section is included to illustrate how a study of the past can help to identify ways of explaining the present. It is about the distribution of a country's population between small, medium and large settlements, which is an important part of its human geography because it has environmental, economic, political and cultural consequences. It might be assumed that highly urbanised countries will be similar in their settlement size distribution, but there are significant differences between them. Table 7.3 shows that the five countries selected have largely similar proportions of their population living in urban areas, so they are all highly urbanised. However, they differ markedly in the proportion of their population living in cities of over one million. Australia in particular stands out as having a high level of urban concentration, with 61% of its population living in five cities of over one million. Correspondingly, it has few medium sized cities compared with the other countries. For example, in the state of New South Wales the Sydney urban area has a population of around five million, while the next largest urban area, Newcastle, has a population of around 500,000. In Western Australia, Perth has a population of around two million, while the next largest city is Bunbury, with only 75,000 people. In comparison, in the state of Kansas in the USA, the largest urban area is Kansas City with around 1.7 million people, but these are spread across the states of Kansas and Missouri. There are then three urban areas within Kansas with populations larger than Bunbury in Western Australia: Wichita (around 650,000), Topeka (around 150,000) and Lawrence (around 95,000). Wichita is an industrial centre, Topeka is the state capital and Lawrence is a university town.

The urban concentration of Australia is a long-standing feature of the country, and has been remarked on since the second half of the nineteenth century.

It has also been regularly criticised as leading to the political dominance of the cities over non-metropolitan Australia; the neglect of rural development and services; and housing, transportation and environmental problems in the big cities.

How can the differences shown in the table, particularly those between Australia and the USA, be explained? The reasons that have been suggested are environmental, historical and political.

1 In Europe and North America, the slow and expensive transportation before the railway age led to the development of a large number of small towns, each serving the rural population of a small area. This was the case with Germany and the UK, and with Canada and the USA for a much shorter period. Australia was colonised by Europeans more than 150 years later than the east coast of North America, so had a very short period of town development before the railway.
2 The colonists of North America and Australia developed export industries that did not use much labour, resulting in relatively small rural populations that could not support the same density of towns as in Europe.
3 Much of Australia has lower rainfall and poorer soils than the USA, particularly in the interior, which again kept rural populations small. Students may be able to explain the climatic differences between the two continental countries by noting their latitudes.
4 The British Government created only four colonies along the east coast of Australia and Tasmania, each of which is now a state with a capital city. Along the much shorter east coast of the USA, there are 16 states, each with a capital city, although not all of them large.
5 Australia's economic role for a long time was as a supplier of commodities to the UK, and manufacturing other than that needed for exports or for goods that were too expensive to import was limited until after 1945. The USA, on the other hand, protected manufacturing after independence and developed a number of manufacturing towns. America also has had a much larger population than Australia and therefore the market size to support local manufacturing.
6 In the USA, government functions such as the political capital and the state university were often allocated to different towns and usually not to the main commercial centre. In Australia, all government functions were located in the main city until relatively recently.
7 When railways were developed in Australia from the mid-nineteenth century onwards, they radiated from the capital city and channelled trade and income into it. In the USA, the railways were constructed west across the continent and created new towns and cities.
8 Australia has had a much more centralised political system than the USA, both as separate British colonies and then as a federation since 1901, and power is concentrated in the capital cities. In the USA, power is much more

dispersed to the states and to local areas. For example, in Australia, royalties from mining go to the state or territory government, whereas in the USA, counties are able to gain revenue from mining and use it for local development. George Wilkinson et al. (2023) argue that this explanation is the dominant one, and that any attempt in Australia to reduce urban primacy and decentralise the urban population will need 'to address the political drivers of primacy, such as fiscal centralization and weak localism' (p. 32).

There are other explanations that could be added, but the list above is enough to illustrate how differences in governmental systems, settlement history and environment can combine to produce different urban patterns. The explanations have focused on the differences between Australia and the USA. Canada is more similar to Australia in population size, environmental limitations and political systems than the USA, but there are still sufficient differences to make Canada a less centralised urban system. This example illustrates the use of time to understand human geographical phenomena, and also the value of a controlled comparison of places, i.e., Australia and the USA, as a method to identify explanatory variables (see Chapter 4).

Models of change over time

Some changes over time in environmental and human phenomena have been generalised into models that describe the stages of change. Two examples are described here.

The mobility transition model

This was developed by the American geographer Wilbur Zelinsky (1921–2013) as a geographical version of the model of the demographic transition. Zelinsky published his ideas in 1971 as 'The hypothesis of the mobility transition', and they have been challenged, revised and revived over the years. He proposed that population mobility followed five stages, the first four of which corresponded with the original four stages in the demographic transition. These are:

1 Phase 1: Pre-modern societies have limited migration, and mobility is mainly circulation associated with the seasonal use of natural resources, or for trade or religious occasions.
2 Phase 2: Early industrialisation societies experience massive rural-urban migration, or movement to colonise land within their country if available. There are also major flows of migrants to other countries, such as from Europe to North America.
3 Phase 3. Late industrialisation societies continue to experience major rural-urban movement, but other forms of mobility decline, with the exception of circulation.

4 Phase 4. Modern societies have declining rural-urban movement, but increasing city-city migration, and increased migration of people from poorer to richer countries. International circulation for work or leisure grows rapidly.
5 Phase 5. Zelinsky added a future superadvanced society phase in which he speculated on trends in mobility in a future that has now arrived. He suggested that there may be a decline in residential migration and some forms of circulation because of the effects of new information and communication technologies, that residential migration would be largely within and between urban areas, that other forms of circulation would increase, and that states may introduce tighter controls on both within country and between country movement.

While Zelinsky's model has been criticised for such things as underestimating the mobility of pre-modern societies, or not fitting the experience of some countries, or for not anticipating the migration of skilled people from the Global South to the Global North, it is still a very useful way of thinking about change. He also contributed some new ideas. For example, he argued that human mobility takes many forms, and distinguished between migration, which he described as 'any permanent or semipermanent change of residence', and circulation. The latter he described as follows:

> Circulation denotes a great variety of movements, usually short-term, repetitive, or cyclical in nature, but all having in common the lack of any declared intention of a permanent or long-lasting change in residence. Under this rubric, one can include such disparate items as weekend or seasonal movements by students; vacation and weekend travel; shopping trips; hospital and church visits; religious pilgrimages; travel to professional and business conventions; trips by government and business executives, salesmen, athletes, migratory farm workers, and the like; social visits; and much seemingly aimless or fun-seeking cruising by wheelborne youngsters.
>
> (Zelinsky, 1971, p. 226)

Zelinsky also anticipated that developments in information and communication technologies might reduce the need to physically move, and his prediction that residential mobility would decline in Phase 3 has been found to be correct in recent studies of some highly urbanised countries. A final contribution was Zelinsky's suggestion that 'states may introduce tighter controls on both within country and between country movement'. Students could look for recent examples of both these situations, especially the latter.

Zelinsky's hypothesis of the mobility transition is an excellent example of a model that describes and to some extent explains change over time, and that also ventures into speculating about the future. It is about one of the most central features of geography, the location and movement of people, and deserves more recognition in school geography.

Vegetation succession

Succession is a common concept in school geography, and refers to the changes over time in the structure and composition of vegetation. Classic accounts of succession argue that these changes are predictable and end in stable climax plant communities, composed of the species best adapted to the climate of a region and resistant to colonisation by other species. Julie Peacock describes the example of succession after the retreat of a glacier and illustrates the interaction between plants and soil:

> the pioneer community is made up of lichens and mosses and other autotrophic species which do not require soil, but can cling to the rock. These species provide food for grazers and decomposers which will then colonise the site. Gradually soil builds up as decomposers break down dead plant and animal matter. The soil and organisms retain water in the ecosystem. Eventually vascular plants will invade the area once the soil is suitable. These will create more litter for the developing soil and more resources for grazing species and decomposers. Pioneer species may be gradually replaced. With a lot more time, conditions will become suitable for woody plants to become established. Woody plants generally require more soil and are slow growing so their establishment is delayed. … Eventually, enough trees will grow to produce a forest canopy and early pioneer tree species … will be replaced by slow growing species. This final community is termed the climax community and is considered to be quite stable.
>
> (Peacock, 2017, p. 288)

At a global scale, these climax communities are the types of vegetation found in the world's biomes. However, vegetation is also influenced by soils, geology, aspect and topography, while succession has been interrupted by storms, disease, fire and human activities, so at a smaller scale it has been difficult in practice to identify discrete and stable climax communities. Studies of vegetation change after the existing vegetation has been destroyed or removed have also found that there are several possible alternative states rather than the one climax community. This growing understanding of the complexity and unpredictability of vegetation change over time, and of the extent of vegetation disturbance, has led most ecologists to abandon the concept of climax communities as a theoretical state that may not exist in reality. The concept of succession is important, however, particularly for understanding ways to restore vegetation on degraded land, but Karen Holl cautions that:

> Scientific studies and observations of recovery in restoration projects make it clear that (1) the recovery process is rarely as predictable as suggested by simple, linear successional models and (2) the shape of the successional trajectory varies greatly among ecosystem types and even across individual sites in a given ecosystem.
>
> (Holl, 2020, p. 54)

Non-linear change

The discussion of vegetation succession has raised the issue of non-linear change. Students are likely to assume that cause and effect relationships over time are linear, in that the same increase in the causal factor (or independent variable) produces the same proportional increase in the effect (or dependent variable), and a larger increase in the causal factor produces a correspondingly larger increase in the effect. With non-linear change the effect may change abruptly and unexpectedly. Two examples of non-linear change that are relevant to school geography are described below.

> Algal blooms and fish kills: Excessive nutrient loading fertilizes freshwater and coastal ecosystems. While small increases in nutrient loading often cause little change in many ecosystems, once a threshold of nutrient loading is achieved, the changes can be abrupt and extensive, creating harmful algal blooms (including blooms of toxic species) and often leading to the domination of the ecosystem by one or a few species. Severe nutrient overloading can lead to the formation of oxygen-depleted zones, killing all animal life.
>
> Fisheries collapses: Fish population collapses have been commonly encountered in both freshwater and marine fisheries. Fish populations are generally able to withstand some level of catch with a relatively small impact on their overall population size. As the catch increases, however, a threshold is reached after which too few adults remain to produce enough offspring to support that level of harvest, and the population may drop abruptly to a much smaller size. For example, the Atlantic cod stocks of the east coast of Newfoundland collapsed in 1992, forcing the closure of the fishery after hundreds of years of exploitation .... Most important, the stocks may take years to recover or not recover at all, even if harvesting is significantly reduced or eliminated entirely.
>
> (Millennium Ecosystem Assessment, 2005, pp. 89–90)

Non-linear change is an increasing phenomenon in physical geography, and in human geography change can also be sudden and unpredictable, because it is produced by human actions and technological innovations. For example, the changes described by Kondratieff could not have been predicted, and continue to alter the geography of the world.

Using change over time to forecast the future

Students can use trends over time to look into the future. For example, they could use changes in environmental indicators such as coastal shoreline retreat, forest cover loss, soil erosion, species decline, fish stocks, ground water levels and atmospheric pollution to identify whether there is cause for concern. Similarly, they could identify recent trends in demographic and economic indicators for their town and project them into the future. If they do not like the result, they could then think about what strategies might produce a different outcome. However, investigations such as these need to be aware of ways in which the future may be different to the past. Global warming will be an additional factor in evaluating environmental changes, and a range of technological, economic and political changes may upset human geographical forecasts of the future.

Conclusion

Scale and time are important concepts in geographical education. All our observations of the world are made at a particular scale, and the scale chosen will influence what we find and how it can be explained, while changing the scale will provide a different perspective and suggest other explanations. Similarly, our observations of the world can be at a single point in time, or over periods of time varying from millions of years in physical geography to hours in human geography. For students, an awareness of time will lead to an awareness of change, and an interest in what the future geography of their world may look like. This may also stimulate them to think about how they can influence that future.

Summary

- Scale in school geography is either cartographic or spatial.
- Spatial scale is the areal size selected to observe the world and is a human construct.
- Geographers use scale as a zoom lens with which to observe the same phenomena at different scales.
- Observations at a large scale conceal variations at smaller scales.
- The choice of scale must match the matter being studied.
- Generalisations made and explanations identified at one scale may not hold at another.
- The choice of scale can affect perceptions of the direction of influences and the appropriate location of actions.
- Networking is a flexible way in which organisations work across scales.
- The concept of time teaches an awareness of change.
- Study of the past can help to identify ways of explaining the present.
- Models of change over time can be used to understand how particular phenomena change and to forecast future change.
- Non-linear change makes such forecasts difficult.
- Change over time can be used to forecast the future, but with difficulty.

How could you use this chapter in teaching?

- In the study of almost any topic, it would be useful to explain to students why the scale of the study has been chosen.
- The effect of scale could be demonstrated by having different groups of students examine the same topic at different scales and then debate their different findings.
- Historical investigations of the local area will illustrate the value of examining a place over time.
- Students could investigate how their local area has been affected by Zelinsky' mobility transition and Kondratieff's cycles and use these models to forecast futures.

These are questions that students could be asked:

- What do geographers mean when they compare the use of scale with a zoom lens?
- Why is it important to think about scale when planning an investigation?
- Which level of scale (i.e., local, national or global) has the most influence on your life?
- What are 'scalar traps', and how can they be avoided?
- How can we use the past to help explain the present?
- Can we use the past to forecast the future?
- What does non-linear change mean, and why is it important?

Note

1 Unfortunately, his views on the economic history of capitalism and the benefits of market-led industrialisation for Russia conflicted with those of Joseph Stalin, and he was executed in 1938, along with about 700,000 others in the Great Purge.

Useful reading

Castree, N. (2015). The anthropocene: A primer for geographers. *Geography*, *100*(2), 66–75. https://doi.org/10.1080/00167487.2015.12093958

Lambert, D., & Morgan, J. (2010). *Teaching geography 11–18: A conceptual approach*. Open University Press (Chapter 7).

References

Brown, J. C., & Purcell, M. (2005). There's nothing inherent about scale: Political ecology, the local trap, and the politics of development in the Brazilian Amazon. *Geoforum*, *36*, 607–624. https://doi.org/10.1016/j.geoforum.2004.09.001

Bulkeley, H. (2005). Reconfiguring environmental governance: Towards a politics of scales and networks. *Political Geography*, *24*, 875–902. https://doi.org/10.1016/j.polgeo.2005.07.002

Burt, T. (2009). Scale: Resolution, analysis and synthesis in physical geography. In N. J. Clifford, S. L. Holloway, S. P. Rice, & G. Valentine (Eds.), *Key concepts in geography* (2nd ed., pp. 199–216). SAGE.

Dicken, P. (2011). *Global shift: Mapping the changing contours of the world economy.* SAGE.

Dittrich, K. (2022). Scale in research on grand challenges. In A. A. Gümüsay, E. Marti, H. Trittin-Ulbrich, & C. Wickert (Eds.), *Organizing for societal grand challenges (Research in the Sociology of Organizations, 79)*. Emerald. https://doi.org/10.1108/S0733-558X20220000079016

Herod, A. (2009). Scale: The local and the global. In N. J. Clifford, S. L. Holloway, S. P. Rice, & G. Valentine (Eds.), *Key concepts in geography* (2nd ed., pp. 217–235). SAGE.

Hoefle, S. W. (2006). Eliminating scale and killing the goose that laid the golden egg? *Transactions of the Institute of British Geographers*, *31*, 238–243. https://doi.org/10.1111/j.1475-5661.2006.00203

Holl, K. D. (2020). *Primer of ecological restoration.* Island Press.

Inkpen, R., & Wilson, G. (2005). *Science, philosophy and physical geography* (2nd ed.). Routledge.

IPCC (2021). Summary for policymakers. In V. Masson-Delmotte, P. Zhai, A. Pirani, S. L. Connors, C. Péan, S. Berger, N. Caud, Y. Chen, L. Goldfarb, M. I. Gomis, M. Huang, K. Leitzell, E. Lonnoy, J. B. R. Matthews, T. K. Maycock, T. Waterfield, O. Yelekçi, R. Yu, & B. Zhou (Eds.), *Climate change 2021: The physical science basis.* Contribution of Working Group I to the Sixth Assessment Report of the Intergovernmental Panel on Climate Change. Cambridge University Press. doi:10.1017/9781009157896.001

Kondratieff, N. D. (1979). The long waves in economic life. *Review (Fernand Braudel Center)*, *2*(4), 519–562. https://www.jstor.org/stable/40240816

Lawson, I. (2017). The Holocene. In J. Holden (Ed.), *An introduction to physical geography and the environment* (4th ed., pp. 108–136). Pearson Education.

Mackinnon, D., & Cumbers, A. (2007). *An introduction to economic geography: Globalization, uneven development and place.* Pearson Education.

Millennium Ecosystem Assessment. (2005). *Ecosystems and human well-being: Synthesis.* Island Press.

Nunn, P. (2018). *The edge of memory: Ancient stories, oral tradition and the post-glacial world.* Bloomsbury Sigma.

Osofsky, H. M., & Levit, J. K. (2008). The Scale of networks?: Local climate change coalitions. *Chicago Journal of International Law*, *8*(2), 409–436. https://chicagounbound.uchicago.edu/cjil/vol8/iss2/4

Peacock, J. (2017). Ecosystem processes. In J. Holden (Ed.), *An introduction to physical geography and the environment* (4th ed., pp. 277–297). Pearson Education.

Pike, A., Rodríguez-Pose, A., & Tomaney, J. (2006). *Local and regional development.* Routledge.

Rawding, C. (2013). *Effective innovation in the secondary geography curriculum: A practical guide.* Routledge.

Steffen W., Richardson, K., Rockström, J., Cornell, S. E., Fetzer, I., Bennett, E. M., Biggs, R., Carpenter, S. R., de Vries, W., de Wit, C. A., Folke, C., Gerten, D.,

Heinke, J., Mace, G. M., Persson, L. M., Ramanathan, V., Reyers, B., & Sörlin, S. (2015). Planetary boundaries: Guiding human development on a changing planet. *Science*, *347*(1259855), 1–10. https://doi.org/10.1126/science.1259855

Wilbanks, T. J., & Kates, R. W. (2009). Global change in local places: How scale matters. *Climate Change*, *43*, 601–628. https://doi.org/10.1023/A:1005418924748

Wilkinson III, G., Haslam McKenzie, F., Bolleter, J., & Hooper, P. (2023). Political centralization, federalism, and urbanization: Evidence from Australia. *Social Science History*, *47*, 11–39. https://doi.org/10.1017/ssh.2022.30

Zelinsky, W. (1971). The hypothesis of the mobility transition. *Geographical Review*, *61*(2), 219–249. https://www.jstor.org/stable/213996

Chapter 8

The evaluative concepts

Sustainability and human wellbeing

The topics studied in school and university geography frequently lead to the identification and examination of a problem or issue. A topic on the coast, for example, may identify coastal erosion as a problem, while a topic on migration may identify the outmigration of young people from rural areas as an issue, and both may lead to an exploration of strategies to address them. However, it is important to be able to explain why situations such as these are problems or issues and not just assume that they are. Coastal erosion and outmigration are not in themselves problems that require responses unless it can be shown that they have consequences that are harmful in some way and are therefore significant. This then raises the question of what criteria to use to evaluate whether these consequences are harmful and in what way. This chapter focuses on two concepts that can be applied to develop criteria for such an evaluation—sustainability and human wellbeing. Both enable students to make judgements about the seriousness of the effects of coastal erosion and outmigration on the environment and on people.

The chapter has a further purpose, which is to raise ethical and values questions for students to consider. David Mitchell and Alexis Stones (2022) argue that the literature on powerful knowledge and powerful thinking neglects to ask the question: powerful for what ends? They suggest that students need separate conceptual tools and reference points to make ethical and values judgements about their thinking. Similarly, Lew Zipin et al. (2015) argue that Young's concept of powerful knowledge overemphasises the cognitive purposes of schooling and marginalises the ethical purposes. This chapter responds by proposing human wellbeing as an appropriate conceptual tool for this purpose.

Sustainability[1]

Sustainability is a widely used concept in geography and geographical education, yet its meaning is often difficult to pin down. In textbooks, it is frequently equated with sustainable development and defined in terms derived from the Bruntland Commission's influential report as 'the ability to meet the needs of the present without compromising the needs of future generations'

DOI: 10.4324/9781003376668-9

(UN World Commission on Environment and Development and Commission for the Future, 1990). This has not helped understanding because, as the words indicate, 'sustainability' and 'sustainable development' are very different concepts. Sustainability is the state or condition of being sustainable, while sustainable development is a process of economic and social change designed to produce an environmentally sustainable economy and a just society. Sustainability is a noun formed from the adjective 'sustainable', which means being able to be maintained or kept going, so something is sustainable if it can be continued into the future. In geography, this meaning can be applied to the environment and to places. Some writers also advocate social, economic, political and cultural forms of sustainability as further applications of the concept. However, these other usages are mostly not about sustaining something into the future, but about some other desirable outcome—such as political democracy, respect for diversity or stable economic growth—for which there are already appropriate terms. Political sustainability, for example, can be achieved by authoritarian regimes such as China, as well as by social democracies, but this is not what is advocated by those who use this term. On the other hand, progress towards environmental sustainability does depend on social, economic, political and cultural factors, such as governmental and corporate transparency, the ability of people to influence public policies through a functioning democracy, cultural attitudes that support sustainability policies, social cohesion and economic equality.

To apply the concept of sustainability to the environment, we must first determine what aspect of the environment should be maintained. If we take a human-centred view, an answer to this question is to start with the idea of our dependence on the environment to support human life and wellbeing. Chapter 6 identified four ways in which the biophysical environment provides this support, and these are:

1 As a source of materials and energy.
2 As a sink into which humans dump wastes.
3 As a provider of ecosystem services that regulate and maintain the environmental conditions on which humans depend.
4 As a spiritual influence on our emotions, health, imaginations and beliefs.

Environmental sustainability then can be defined as the maintenance of these environmental functions or services, because they are vital to human wellbeing and even survival (Ekins, 2000; Goodland, 1995). The advantage of focusing on this way of defining sustainability is that it is then possible to develop statements about what sustainability means for each function. For some it is also possible to identify proxy measures which can be used to determine the sustainability of that function in a specific context or place. The sustainability of other functions, however, can only be determined by subjective judgements, which depend on individual ideologies, worldviews and values. The meanings of sustainability for the four environmental functions are described below.

The use of environmental materials and energy

The sustainability of our use of materials and energy can be assessed by two principles.

Renewable resources should be extracted at or below their rates of renewal, and in ways that do not reduce the productive capacity of the environment.

This principle has been used for some time in the management of fisheries, forests and groundwater. For example, if fish are caught at a faster rate than they can breed, or the marine environment in which they breed is damaged by fishing methods, then the stock of fish will decline and the fishery will not be sustainable. The near disappearance of the cod fishery off the coast of Canada in the 1990s is a salutary example of the consequences of ignoring this principle (Sale, 2012). Good fishery management aims to maintain viable stocks of fish and to minimise the impacts of fishing on the productivity of marine ecosystems. This is done by careful monitoring of fish stocks, control of the quantity of fish that can be caught, regulation of fishing methods and the designation of marine areas as reserves from which fishing is excluded. Similar problems can occur with the unsustainable extraction of water from rivers and ground water, with the Aral Sea an extreme example of over-extraction. Other examples include the growing problems resulting from excessive ground water extraction in the High Plains of the USA and the North China Plain, and surface water extraction from the Colorado River in the USA and the Murray-Darling river system in Australia.

The second half of the principle ('in ways that do not reduce the productive capacity of the environment') applies not only to marine ecosystems but also to the effects on agricultural and forestry production of soil erosion, compaction, acidification and salinisation resulting from vegetation removal, cultivation, road construction, irrigation and fertilisation. These types of land degradation can reduce the productive capacity of the environment by lowering yields and sometimes by removing land from production altogether. The principle also applies to the effects of soil erosion and chemical residues on terrestrial and marine aquatic environments and of animal emissions of methane on climate change. Ensuring the sustainability of renewable resources consequently requires more than avoiding over-extraction; the productivity and stability of the environment must also be maintained.

The sustainability of a renewable resource can be measured, although not always precisely, by indicators such as trends in the stock of fish or timber, the depth of water tables, the rate of soil erosion and trends in soil acidification or the area of land lost to salinisation.

Non-renewable resources should not be extracted in ways that damage other environmental functions.

Non-renewable resources differ from renewable ones in that when they become scarce their price rises, and more effort is made to recycle material, locate new reserves and/or develop substitutes. The market should therefore find a solution, although government action may be necessary if the market fails and serious shortages of critical substances emerge. On the other hand, mining activities that degrade productive environments through pollution or the depletion of groundwater resources, or add to carbon emissions, are a significant threat to environmental sustainability.

Students may assume that the use of a non-renewable resource cannot be made sustainable, because it cannot be renewed by natural processes. However, the use of a non-renewable resource can be partly sustained through reuse and recycling, and future technology may make it possible to recycle up to 90% of some minerals. Another way in which a non-renewable resource can be made sustainable is through the conversion of some of its value into financial capital. For example, the governments of Norway and Zambia have both established sovereign funds from the taxes and royalties their countries have earned from mining, and when such income is invested and managed carefully, the financial capital becomes a continuing resource that does not get depleted.

Waste disposal

This function is the capacity of the environment to safely absorb, through breakdown, recycling or storage, the wastes and pollution produced by production and human life. Its sustainability can be expressed by two principles.

Biodegradable wastes should not be added to the environment in ways that prevent them from being broken down and safely recycled or stored, or reduce the productive capacity of the environment, or threaten human health.

This principle refers to the ability of microorganisms to break down biodegradable materials and either store the resulting elements or make them available for plant growth. However, if discharges of organic matter from agricultural land, sewage plants or agro-processing factories are too large, they can cause the growth of toxic algae in inland and coastal waters. This is harmful for people and animals and, as it consumes oxygen, it also makes the environment uninhabitable for aquatic life. If this principle is not followed, a water resource may become less productive or even unusable because of eutrophication and consequently less able to support human welfare into the future.

Non-biodegradable wastes should not be added to the environment at levels that threaten human health or other environmental functions.

This principle is about wastes that environmental processes are unable to make safe. If the principle is not followed, the provision of environmental resources could be threatened by toxic chemicals that make land or water resources unusable, or kill marine and aquatic life. Human health could also be threatened. The principle underlies environmental regulations that ban or limit the discharge of toxic substances into the environment, with the aim of keeping their concentrations below the levels that threaten the health of humans and other living bodies.

The sustainability of both types of waste disposal can be measured by a range of indicators, including the levels or concentrations of nitrates in rivers and groundwater, pesticides in the environment, oxygen in aquatic environments, pollutants in the atmosphere and heavy metals in soils.

Ecosystem services

These are environmental services that support human life and wellbeing without requiring human action, such as those that maintain stable climates or pollinate plants. Their sustainability can be expressed by this principle:

The ecosystem services that regulate and maintain the environmental conditions on which humans depend should be maintained.

The sustainability of some of these services can be measured by indicators such as trends in the size of the hole in the ozone layer, local and global temperatures, rainfall, hazards such as wildfires and cyclones, and bee populations. However, scientific understanding of some of these services is incomplete, and views on their sustainability are often subjective and divided. The maintenance of a stable climate is the most important of them and is currently threatened by global warming.

Spiritual functions

There is a considerable difference between the ecosystem services of the environment and what is here called the environment's spiritual functions. The former exist independently of human thought, can be studied scientifically and are the same for all people. The spiritual functions only exist because of human thought, as they involve our emotions, imaginations and beliefs, and these vary from person to person, culture to culture and over time. Their sustainability can be expressed by this principle:

The recreational, psychological, aesthetic and spiritual value of environments for people should be maintained.

The spiritual functions of the environment cannot be measured objectively because they involve human perceptions and beliefs, so it is difficult to reach agreement on whether and how this function should be maintained. The principle enables teachers to explore students' enjoyment of environments, appreciation of their beauty and their feelings of wonder and awe.

Two more sustainability principles could also be discussed with students, and these are:

An assessment of the environmental sustainability of a place or country must take into account the effects of its production and consumption on the environmental sustainability of other places.

Global sustainability depends on an equitable sharing of global environmental services.

These two principles are essential to a global view of sustainability. The first, known as the principle of trans-frontier responsibility, says that cities, regions and countries cannot be considered sustainable if their production processes and consumption patterns are responsible for environmental unsustainability somewhere else. This would be if they imported commodities from other places that were produced by unsustainable practices, as in the case of some timber imports into high-income countries, or if they exported wastes by water or through the atmosphere to other places.[2] The second principle reminds students that any strategy to manage the total pressure of humans on global environmental functions requires an equitable sharing of these functions. This is the same as the intra-generational principle associated with the Bruntland Commission's definition of sustainable development.

Defining sustainability as the maintenance of the four environmental functions leads to clearer ways for students to assess whether something is sustainable than describing it as ensuring the ability of future generations to meet their own needs from the environmental resources they inherit, as is implied in the World Commission on Environment and Development's description of sustainable development. This is because we do not know the needs of future generations or the technologies that may be developed to help them meet them, so we cannot determine what environmental resources they will require other than through subjective judgments. On the other hand, for many of the environmental functions and services discussed above, although not for the more subjective ones, we can identify whether they are being sustained at present, and therefore whether their capacity to support human life and wellbeing is likely to continue into the future. This way of thinking about sustainability has the added advantage of emphasising the present, as many sustainability issues are affecting current generations and not only future ones.

It is also preferable to the Ecological Footprint (EF), which is widely used in education to evaluate the sustainability of people's ways of living. The EF is a measure of how much of the Earth's environment (in land area) a person, a city or a nation is estimated to be using to produce what they consume and to dispose of wastes, including carbon emissions. However, it doesn't measure the sustainability or unsustainability of that production and waste disposal, and it has been widely criticised for basic methodological weaknesses, for the way it aggregates quite different environmental problems, and for having no policy value. The discussion about layers of explanation in Chapter 2 illustrates another problem with the EF. My personal EF is 5.1 global hectares[3] and is below the Australian average of 7.3 global hectares because I am retired and don't travel as much as an employed person. Even so, if everyone lived like me the footprint calculator estimates that we would need three Earths. Most of my footprint (64%) represents the land required to offset my carbon dioxide emissions. There is very little I can do personally to further reduce these emissions, as I already have solar panels on my house to generate electricity. On the other hand, if all power in Australia was generated without burning fossil fuels, and all transport was similarly powered, my footprint would be only 1.8 global hectares, and if everyone lived like me then we would only need one Earth instead of three. However, I can only achieve this desirable situation if industries and governments take the appropriate actions, and the footprint calculator doesn't enable me to measure what my footprint would be if they did. My criticism of the EF concept is that it puts all the emphasis on individual actions, and not on the deeper layers of explanation which individuals have no ability to change. This can leave people feeling powerless and unable to see any solutions.

Applying the principles can be complex. An example is the strategy of reducing carbon emissions by replacing fossil fuels with renewable biofuels, adopted by some governments in response to public pressure. This may conflict with environmental sustainability or human wellbeing, or both, if it requires the clearing of land or produces soil degradation, displaces poor farm families or reduces food supplies and raises food prices by diverting food crops to biofuel production. For example, the strategy has been blamed for provided an incentive for deforestation in the Amazon Basin, as discussed in Chapter 10.

Sustainability as a contested concept

Sustainability can be a highly contested concept, in two main ways. First, for all of the functions of the environment there can be disagreement over whether a situation is serious enough to require action. Some may even deny that a problem exists, despite the scientific evidence. This attitude is often based on a dislike of government regulation of business, and a belief that economic and social considerations, like the generation of income and employment, should always have precedence over environmental sustainability. Others may use evidence of a problem to advocate greater regulation, or changes in our economy, or simpler ways of living, because of their dislike of capitalism and consumerism.

Second, there are different views on how to achieve sustainability. For example, to ensure the sustainability of a renewable resource, some prefer methods that reduce the use of the resource, such as restrictions on fishing, or bans on water use, or recycling paper and cardboard to reduce the need to plant more trees. Others want methods that increase the output of the resource, such as planting trees, improving soil fertility, damming rivers, farming fish (aquaculture) or making fresh water by desalination.

People's views on how to achieve sustainability reflect their environmental ideology or worldview, even though they may be quite unaware of what this is. These range from an ecocentric (environment-centred) view that humans are a part of nature and must manage the environment for the welfare of all life forms, to an anthropocentric (human-centred) view that humans are separate from nature and that the environment is there to provide for human needs. The latter tends to see technology as the solution to sustainability problems, while the former advocates working with nature and changes in people's ways of living. These worldviews are another good topic for student discussion.

Sustainability is sometimes criticised as being opposed to any development that produces environmental change. This is a misunderstanding of the concept, as sustainability is not about preventing environmental change, but about ensuring that change maintains the environmental services that support human life and human welfare. Humans have a long history of transforming their environments in ways that have enabled the world to support more people at higher levels of welfare. If we had remained hunter-gatherers and not cleared land for farming, the world might support about only 100 million people and would not have developed urban civilisations. If modern agricultural technologies had not been developed the world might support only about three billion people, instead of the present eight billion. Our way of life has been made possible by changes to the environment. Sustainability is about ensuring that these changes do not threaten important environmental services, through actions like carefully managing renewable resources, preventing soil degradation, restricting climate change and preserving genetic resources.

Students need to understand the causes of the unsustainability of an environmental function or service before they can debate solutions, yet these causes are also likely to be contested. To find them, students should follow the causal interconnections between the immediate actions that are producing the problem (known as the proximate causes) and the factors that are causing these actions (known as the underlying or indirect causes). There may be several layers to an explanation, as discussed for land degradation and climate change in Chapter 6.

The contestability of some aspects of sustainability gives students the opportunity to explore topics and issues for which there are no simple answers, and where they have to evaluate conflicting opinions and sometimes conflicting evidence. They can debate some big issues and learn to look critically at claims made by protagonists on both sides of an environmental argument. This is an important preparation for citizenship.

Sustainability and physical geography

Sustainability provides an excellent context for demonstrating the importance of physical geography to students, because to understand the environmental functions discussed above, and the conditions for their sustainability, students need some understanding of physical processes. Some examples are outlined below:

- To understand sustainability issues for water, students need some knowledge of the water cycle, and of the effects of precipitation, runoff, evapotranspiration and infiltration on soil moisture and surface water.
- To understand sustainability issues for soils, students should know something about rates of soil formation and erosion; the effects of cultivation on soil structure, plant nutrients and organic matter; and the causes of soil

salinity and acidification. These are much more relevant to an understanding of sustainability (and of food and fibre production) than the common study of soil types, and much more interesting for students.
- To understand the sustainability of biodegradable wastes, students need some knowledge of the processes that break down and recycle or store these wastes and of the causes of eutrophication.
- To understand some of the problems with the disposal of non-biodegradable wastes, students need some knowledge of the movement of groundwater.
- To appreciate the importance of ecosystem services, students need to know something about how these services work.
- To identify ways of protecting valued landscapes, students need some knowledge of the processes, like vegetation degradation and soil erosion, that damage them.

These are just a few examples, but they suggest a way of orienting the teaching of physical geography towards an understanding of a range of practical and contemporary issues. This is in keeping with the advice of Rachel Atherton (2009) to teach a physical geography that is about how natural processes relate to human activity, as a way of engaging students in topics they don't always find interesting.

Sustainability and place

The concept of place provides a particularly geographical way of exploring sustainability issues. This is partly because most environmental sustainability issues are local. In the classification of the ways that the environment supports human life and wellbeing described earlier, most of the examples are local in their spatial extent and are caused by local actions and require local solutions. Sustainability issues such as soil erosion, loss of biodiversity, declining water tables and loss of environmental amenity are local, although they may have non-local causes. While they may cover a wide area, they have little or no effect on other places. Atmospheric pollution, on the other hand, has local origins, but can spread to quite distant places. There are only a few truly global issues, produced by global environmental processes, but even a clearly global one such as climate change is produced by the use of fossil fuels in individual places, and different places have different sources of energy, different consumption patterns and different contributions to greenhouse gases. Strategies to reduce carbon dioxide emissions must take account of these differences. Eric Pawson aptly describes this geographical contribution to ESD as 'countering grand narratives with local encounters' and identifies strategies being tried by many cities to mitigate and adapt to climate change as examples (Pawson, 2015, p. 309). Economic and social sustainability issues are also often local, and some of the action and pressure to respond to them operates at a local level, by citizens and local governments. These are activities in which students can become involved themselves.

Applying the sustainability principles

In studying sustainability issues, it is vitally important that students examine situations where effective actions are being taken to improve sustainability and are not overwhelmed with negative examples. These may best be studied at a local or regional level, and Casinader and Kidman (2018) make a strong case for the value of geographical inquiry through local fieldwork in helping students to understand the complex human causes of environmental problems and the variety of ways of managing them. Examples will vary according to location, but could include:

- Agricultural methods that improve farm soils and water resources.
- Rehabilitation of mining land.
- Projects to improve vegetation cover and water retention in urban areas (see Chapter 6).
- Construction of wetlands to retain and purify water.
- House design to reduce energy consumption.
- Methods for supplying renewable energy when the wind isn't blowing and the sun isn't shining.
- Ways to reduce carbon emissions from transportation.
- Ways to make tourism less unsustainable. In my experience, students generally think of sustainable tourism as being about the management of the environmental, economic and social impacts of tourism on the destinations and neglect to think about the transportation used to get to them and return.
- Strategies to reduce the environmental impacts of clothing, a major problem that is not well recognised (see Goodall, 2012).
- Exploring the concept of a circular economy and its application in local recycling programs.
- Projects to conserve biodiversity.

Why does biodiversity matter?

Biodiversity is the variety of life on Earth. This life is important because it provides humans with the resources and services described in Chapter 6, but there is another reason. This is that variety makes ecosystems more able to cope with changes in their environment. For example, in a forest with a variety of tree species a new pest or disease, or a changed climate, may only weaken some species, as others may be more resistant to the pest or disease, or more suited to the changed climate. More generally, natural ecosystems usually contain several species that can perform the same function, so the loss of a few will not degrade the goods and services that these ecosystems provide. However, accumulated losses may remove all members of a functional group, and widespread species loss may produce unexpected changes. Peter Sale writes:

> We have growing evidence that ecological change can be abrupt rather than gradual, that thresholds and tipping points really do exist. It is possible that the loss of certain species will increase the likelihood of loss of certain other species strongly dependent upon them, so that the rate of loss spirals up to much higher rates than at present. It is also possible that the growing stress on the environment caused by our activities will accelerate species loss. In both cases, thresholds and tipping points are likely to loom up unexpectedly. So, while we know that species are going extinct at a fast rate, what we do not know at present is how far down this path of lost species we can go before the situation becomes critical and we find ourselves at the threshold of a nightmare. I would prefer not to find out.
>
> (Sale, 2011, pp. 225–226)

A reduction in the variety of species consequently threatens the ability of an ecosystem to continue to provide services for humans. This view of the value of biodiversity provides one argument for protecting animal species, because of their role in maintaining functioning ecosystems. There are, of course, also strong ethical arguments.

A similar problem has been experienced at times with the monoculture of commercial crops, where a new disease can cause widespread damage. Intercropping, the growing of two or more spatially intermingled crops, has been shown to have other benefits. For example, a study in China reported that:

> Using four long-term (10–16 years) experiments on soils of differing fertility, we found that grain yields in intercropped systems were on average 22% greater than in matched monocultures and had greater year-to-year stability. Moreover, relative to monocultures, yield benefits of intercropping increased through time, suggesting that intercropping may increase soil fertility via observed increases in soil organic matter, total nitrogen and macro-aggregates when comparing intercropped with monoculture soils.
>
> (Li et al., 2021, p. 943)

Biodiversity has many benefits.

What can be learned about sustainability from Australian Aboriginal cultures?

Aboriginal people have been living sustainably in Australia for ten of thousands of years, and the environment observed by the early European explorers and colonists was in good condition. Aboriginal people had changed that environment through their practices of controlled burning, but had not degraded it. They had a variety of methods to maintain sustainability, such as prohibitions on catching animals when they are breeding, prohibitions on harvesting plants that are seeding, controlled burning, population control and planned mobility to allow plant and animal resources to regenerate. Some of these are similar to the methods adopted today to maintain fish stocks, ground water resources and timber supplies. Contemporary Australia can learn something from these practices, and a good example is cultural burning. Done at the right time of the year and with the right methods, Aboriginal burning practices produce a cool burn which reduces fuel load, prevents large and destructive fires and the carbon they emit and regenerates grasses for wildlife. In Chapter 6, the use of burning to maintain a mosaic of types of vegetation was described. Recent research has shown that across Southeast Australia these methods maintained grass-dominated ecosystems which had only 5–15% tree cover. The dispossession, death and removal of Aboriginal populations from their lands stopped this burning, resulting in an increase in shrubby cover and a decline in grassy understories in today's forest/woodland areas, changes in fuel availability and increasingly severe bushfires. The researchers argue that:

> Australia's current fire crisis can therefore trace its origins back to the colonial suppression of Indigenous cultural burning and subsequent attempts to suppress landscape fire. Our research informs debates about the role of disrupted Indigenous fire management in other landscapes; for example, there is growing acceptance that suppression of Indigenous fires in North America has contributed to uncontrollable wildfires associated with the accumulation of woody biomass in flammable forests.
>
> (Maraini et al., 2022, p. 298)

A rather different contribution of Aboriginal culture to the present is the ways of thinking that underlie their sustainability practices. These are briefly outlined below.

1 Aboriginal people do not perceive themselves as separate from nature, but as integrated with and closely related to the non-human world, and they do not see a difference between culture and nature.
2 Aboriginal culture has a holistic view of the environment and the interconnections between its elements, including the reciprocal interconnections of humans with the animate and inanimate world around them. This is an understanding of the mutual interdependence that is advocated as essential for a transition to sustainability.
3 Traditional Aboriginal culture is non-materialistic, with considerable emphasis on social relationships and cultural activities. For sustainability, it is argued that societies need to shift from constantly increasing material consumption to a greater emphasis on activities that do not use large quantities of materials and energy, such as cultural activities, recreation and social relationships.
4 Aboriginal culture teaches respect for the environment, and that our actions must do no unnecessary harm.

5 Aboriginal culture teaches that people have a responsibility to care for the sustainability of their world and to pass it on to future generations in good condition. Aboriginal Australians, for example, perceive themselves as custodians of the environments in which they live and with which they are interconnected. As Deborah Rose describes:

> Aboriginal people's land management practices, especially their skilled and detailed use of fire, were responsible for the long-term productivity and biodiversity of this continent. In addition to fire, other practices include selective harvesting, the extensive organisation of sanctuaries, and the promotion of regeneration of plants and animals. Organised on a country by country basis, but with mutual responsibilities being shared along Dreaming tracks, and through trade, marriage, and other social/ritual relationships, management of the life of the country constitutes one of Aboriginal people's strongest and deepest purposes in life …
>
> (Rose, 1996, p. 10)

These ways of thinking are shared by many Indigenous groups around the world. As an example from a very different society, a study of the views of Skolt Sami people in Finland of their obligations to nature found that the most common response involved respect for the environment. The researcher described their attitude as being 'that human beings should be humble and thankful towards nature, accept its superiority, and not try to subordinate it' (Itkonen, 2022, p. 299). These are ways of thinking that non-Indigenous people will have to at least partially adopt if the world is to become more sustainable.

The sustainability of places

Geographers are sometimes interested in the sustainability of places. There is not the space to explore this topic here, but there are many places whose sustainability is threatened by the loss of economic activities, employment and population. These include mining settlements where the mine has closed, towns whose main industry has relocated and small rural settlements serving a declining farm population or unable to compete with larger towns. Their sustainability is also threatened by the loss of young people through outmigration, ageing populations, declining services and low incomes. Places like these will be discussed in Chapter 11.

Human wellbeing

The rest of this chapter discusses human wellbeing as a second concept that can be used to evaluate the significance of whatever students are learning or discovering. The results of this evaluation can then be examined for their ethical consequences. Elizabeth Olson and Andrew Sayer (2009) argue that this ethical critique requires a norm or standard against which to judge what is good and what is not, and they propose human wellbeing for this normative purpose. So what is wellbeing? It is a complex concept with a wide range of possible meanings. For some it is the ability to satisfy the basic needs of food, water and shelter, a definition that applies mainly to low-income countries. A second viewpoint is that wellbeing is about people's personal satisfaction with the quality of their lives and their feelings of physical and mental wellness. This concept provides for individual and group differences about what constitutes a satisfactory life. A third and much broader idea is that wellbeing is about people having the capabilities and freedom to live the life they choose and find valuable. This concept differs from the previous one in that some of it is about the context in which people live and which constrains or facilitates their ability to live the life they choose, such as the educational opportunities available to them. These three views of wellbeing have been described as follows:

> Objective wellbeing is concerned with the material conditions of a person's life, often represented by wealth indicators of poverty. Subjective wellbeing is concerned with self-evaluation of personal circumstances. Examples of subjective wellbeing measurement include the Satisfaction with Life Scale, a five-question research instrument where respondents self-report their satisfaction with life as a whole. Thirdly, relational wellbeing, based on the capabilities approach of economist Amartya Sen, concerns the opportunities available to a person, recognising that individual wellbeing is pursued in relation to other people.
>
> (Loveridge et al., 2020, p. 462)

These conceptions of wellbeing range from very basic ones that are mostly applicable to lower-income countries to more complex ones that can be applied to any country or region. They have in common the belief that economic indicators alone are unsatisfactory measures of human development and wellbeing, as these tend to be about factors that might produce higher wellbeing, such as income, and are not indicators of the wellbeing itself, such as good health. These different ways of conceptualising human wellbeing suggest that when students are evaluating people's wellbeing they should think of indicators such as

- Food security.
- Housing.
- Safe water.
- Adequate financial and other resources.
- Personal physical and mental health.
- Personal safety, both at home (from domestic violence and crime) and in public spaces.
- Access to essential services, such as health.
- Access to education.

- Satisfaction with one's life. This is a measure of the elusive concept of happiness, or subjective wellbeing as it is often termed in the economic literature.[4]
- The capabilities and freedom to live the life one chooses.

What some of these mean in practice will vary between societies. For example, geographers working in rural places in regions such as Southeast Asia and the Pacific Islands have found a preference for forms of development that are responsive to local aspirations and conditions, recognise the importance of customary social relations and promote self-determination and independence. Such development pathways 'may better serve the place-based needs and desires of people who are seeking to maintain and enhance a way of life' (Curry, 2009, p. 418). Similarly, for Aboriginal people in Australia, wellbeing may mean the ability to participate in cultural activities, communicate in their own language where this still exists, manage their ancestral lands and have control over decisions that affect their lives.

How can students apply these concepts of human wellbeing to evaluate what they learn? One way would be to evaluate the social and economic conditions of the people who are the subjects in whatever topic they are studying, using any of the measures above and others that they might think of, and discuss the implications of what they find out. For example, has wellbeing been increasing or declining? Is the wellbeing of people adequate as judged by their expectations? Are there inequalities in wellbeing within the population that seem unjust? Is wellbeing threatened by current or proposed changes?

A second way is based on the capabilities approach developed both together and separately by Amartya Sen and Marta Nussbaum, as outlined in this passage.

> Sen's conception of capabilities emphasises the freedom that individuals should have to make choices in their life, and is built upon concerns about different variants of inequality. His capabilities approach thus reflects not only a consideration of the perpetuation of class inequalities, but also the processes through which gender inequality is reinforced by both men and women. A person's capability is determined by the real opportunities to have and be things ('functionings') that they have reason to value, such as health, security and respect, which might be constrained by social or personal circumstances. Poverty becomes defined as capability deprivation, or the inability to achieve a collection of functionings.
>
> (Olson and Sayer, 2009, p. 190)

Using this approach, students could identify situations and contexts that may restrict people's ability to improve their wellbeing, such as:

- Patriarchal cultures that limit the opportunities for girls and women.
- Discrimination against minority groups.
- Racial discrimination.
- Lack of services to enable people with disabilities to participate in education and employment.
- Poverty.
- Low quality of education and health provision in disadvantaged areas.
- Exploitative employment conditions.
- High levels of crime.
- The absence of the political freedom to advocate for change.
- Weak or non-existent democratic institutions.
- The influence of powerful corporations over government policies and actions.
- The lack of care for those who need it at different stages of life. Virginia Lawson argues that '[d]espite the centrality of care to human well-being, care is still marginalized in geographical theory, in public and political discourse and in our economic lives' (Lawson, 2009, p. 210).
- The safety of the environment in which people live (e.g. absence of harmful pollution).
- The 'greenness' of the environment in which people live, which influences their physical and mental health, as discussed in Chapter 6.

These shift the focus away from the individual to the contexts that have an influence on their ability to improve their lives.

Does subjective wellbeing have a geography?

Place is another context that can influence people's sense of wellbeing. There have been studies of international, intra-national, inter-city and local differences in subjective wellbeing, as typically measured by people's perceptions of their satisfaction with their life. These studies attempt to identify if, after controlling for variables such as age, income and education, there is a residual that represents the effect of place. An example comes from the UK, where a study by the New Economics Foundation (Abdullah and Shah, 2012) analysed the answers to four questions on subjective wellbeing from a national survey conducted in 2011. After statistically controlling for the effect of multiple deprivation, based on measures of income, employment, education, health, crime, housing and living environment, there were areas with higher or lower wellbeing than might be expected. The Southwest of England, Scotland and Northern Ireland were higher than expected, and London was lower. Subjective wellbeing did not have the same spatial pattern as income, and seemed to have a coastal orientation.

Other studies report the following:

- Rural residents tend to report higher life satisfaction than urban people, although a study finds the opposite in poorer countries.
- People in large dense cities tend to report lower life satisfaction than those living in smaller cities and towns, and it appears that continued agglomeration reduces people's quality of life.
- Within cities access to local facilities and services, and the number and variety of these facilities and services, improves people's life satisfaction, as does access to green space.
- Within cities areas with higher levels of air pollution, or higher levels of noise, have lower levels of life satisfaction.

There is clearly a role here for planning and regulation in improving people's subjective life satisfaction.

Wellbeing and the environment

There is a close relationship between wellbeing and the environment. As noted earlier, wellbeing is increased when people live and work in environments that are green and unpolluted. However, not only are many places not green and unpolluted, but also studies find that environmental pollution disproportionally affects minority groups and low-income populations. In the USA, for example, waste dumps are predominantly located in areas where the majority population is African American. Is this ethical? Students could investigate degraded or polluted areas near them to see if they are predominantly areas with low-income or minority populations and debate the ethics of what they find.

The effect of climate change on human health and wellbeing is a major contemporary concern. An IPCC report outlines the issues in this passage:

> Changes in the magnitude, frequency and intensity of extreme climate events (e.g., storms, floods, wildfires, heatwaves and dust storms) will expose people to increased risks of climate-sensitive illnesses and injuries and, in the worst cases, higher mortality rates. Increased risks for mental health and well-being are associated with changes caused by the impacts of climate change on climate-sensitive health outcomes and systems. Higher temperatures and changing geographical and seasonal precipitation patterns will facilitate the spread of mosquito- and tick-borne diseases, such as Lyme disease and dengue fever, and water- and food-borne diseases. An increase in the frequency of extreme heat events will exacerbate health risks associated with cardiovascular disease and affect access to freshwater in multiple regions, impairing agricultural productivity and increasing food insecurity, undernutrition and poverty in low-income areas.
>
> (IPCC, 2022, p. 1126)

Whose wellbeing?

Students are likely to discover that reaching a decision about wellbeing is not at all straightforward. An example is the increasing trend in holiday and tourist places around the world for accommodation that was previously rented to local people to now be rented to visitors at much higher prices, through organisations such as Airbnb. The previous residents are forced to either pay more to compete or to move to more distant locations not preferred by tourists. Their wellbeing is clearly reduced, but at the same time, the wellbeing of many local people is dependent on tourism and therefore on tourists finding the accommodation they want. Tourist expenditure in inner city areas has also helped their revitalisation, from which local people have benefitted. The dilemma facing local authorities in these places is how to minimise the negative effects of short-term tourist rentals while retaining or increasing the benefits.[5]

A second example is what has been described as the 'urban paradox'. This is the tendency noted earlier for average subjective wellbeing to decline as the size of cities in high-income countries increases, despite their apparent economic success. The term 'urban paradox' was proposed by the New Zealand geographer Philip Morrison, and his explanation is summarised by Fredrik Carlsen and Stefan Leknes as follows:

> Morrison suggests that the paradox is due to a composition effect. Since large cities allow the well-educated to receive high wages, there is high demand for tertiary sector services, often supplied by low-paid, low-educated workers. This causes in-migration, which drives up house prices and contributes to segregated cities, where the well-off can afford attractive neighbourhoods closer to work, whereas others are forced to commute long distances, which reduces time for family and leisure. Wage inequalities, neighbourhood segregation and commuting combine to create large gaps in subjective well-being. Although many people with a high education level and high income are happy with life in big cities, average subjective well-being is pushed down by the numerically dominant low-educated and less happy inhabitants.
>
> (Carlsen and Leknes, 2022, p. 2177)

Other topics that could be used to stimulate discussion of 'whose wellbeing?' include:

- The redevelopment of old dockland districts.
- The gentrification of old inner city districts.
- The construction of transport corridors that divide existing communities.
- The creation of wildlife reserves that exclude farming by local people.
- Loss of farmland through urbanisation.

Conclusion

This chapter has discussed how sustainability and human wellbeing are concepts that can be used to evaluate the significance of what students are learning or finding out. If they conclude that there is a significant problem, in that there is an appreciable impact on sustainability or human wellbeing, they can then evaluate the ethics of these impacts, using the approaches suggested in the chapter. In doing this, they need to make explicit the reasons for their ethical judgements and not just assume a view of what is just and what is not.

Summary

- The concepts of sustainability and human wellbeing can be used to evaluate whether the subject of student study is something that has unwanted consequences and is therefore a significant issue.
- They can also be used to make decisions about the ethics of what they find out.
- Sustainability is the maintenance of the environmental functions or services that support human life and wellbeing.
- Principles can be developed that state what sustainability means for each of these functions, and the sustainability of some of them can be measured.
- Cities, regions and countries cannot be considered sustainable if their production processes and consumption patterns are responsible for environmental unsustainability somewhere else.
- Global sustainability depends on an equitable sharing of global environmental services.
- Sustainability is a contested concept.
- Explanations of unsustainability progress from the immediate actions that are causing the problem to the underlying causes of these actions.
- Sustainability issues have local causes, even when they are a global problem.
- The concept of sustainability can also be applied to places.
- Australian Aboriginal ways of thinking about their relationship with the environment provide examples of the shifts in thinking that may be needed by non-Indigenous societies.
- Wellbeing can be conceptualised as objective, subjective or relational.
- Place has an influence on people's sense of wellbeing.
- The environment has an influence on wellbeing.
- Using wellbeing to make judgements about the ethics of a situation is rarely straightforward, as different groups in a population will be affected differently.

How could you use this chapter in teaching?

- Ideas about sustainability should be introduced progressively over the school years. Principles 1, 3 and 4 can be introduced in the upper primary years and the rest in secondary school. All of the principles should be taught in conjunction with the physical geography needed to explain them.
- Ethical questions can also be raised in primary school, as whether something is fair. In secondary school, students should be helped to identify issues that require ethical judgements and challenged to explain and justify their conclusions.
- Sustainability and wellbeing are evaluative concepts that can be applied to many of the topics studied in school geography. Like the other core concepts, they should be taught through the substantive topics in the geography curriculum and not on their own.

Notes

1 The discussion of sustainability in this chapter has been adapted from Maude (2014a, 2014b, 2023).
2 For some examples, see Maude (2023).
3 As calculated by the Ecological Footprint Calculator of the Global Footprint Network (https://www.footprintcalculator.org).
4 For a criticism of happiness as a measure of wellbeing, see Chapter 4 in Skidelsky and Skidelsky (2013).
5 For a discussion of this topic, see Nieuwland and van Melik (2020).

Useful reading

Kirby, A. (2014). Adapting cities, adapting the curriculum. *Geography*, *99*(2), 90–98. https://doi.org/10.1080/00167487.2014.12094399

Maude, A. (2014a) A sustainable view of sustainability? *Geography*, *99*(1), 47–50. https://doi.org/10.1080/00167487.2014.12094391

Maude, A. (2023). Using geography's conceptual ways of thinking to teach about sustainable development. *International Research in Geographical and Environmental Education*. https://doi.org/10.1080/10382046.2022.2079407

Mitchell, D., & Stones, A. (2022). Disciplinary knowledge for what ends? The values dimension in curriculum research in a time of environmental crisis. *London Review of Education*, *20*(1), 23. https://doi.org/10.14324/LRE.20.1.23

Walshe, N., & Perry, J. (2022). Transformative geography education: developing eco-capabilities for a flourishing and sustainable future. *Teaching Geography*, *47*(3), 94–97. https://portal.geography.org.uk/journal/view/J003544

References

Abdallah, S., & Shah, S. (2012). *Well-being patterns uncovered: An analysis of UK data.* New Economics Foundation. https://neweconomics.org/uploads/files/60770a0ad7bce041e2_gcm6b0nfk.pdf

Atherton, R. (2009). Living with natural processes—Physical geography and the human impact on the environment. In D. Mitchell (Ed.), *Living geography: Exciting futures for teachers and students* (pp. 93–112). Chris Kington.

Carlsen, F., & Leknes, S. (2022). For whom are cities good places to live? *Regional Studies*, *56*, 2177–2190. https://doi.org/10.1080/00343404.2022.2046724

Casinader, N., & Kidman, G. (2018). Fieldwork, sustainability, and environmental education: The centrality of geographical inquiry. *Australian Journal of Environmental Education*, *34*(1), 1–17. https://doi.org/10.1017/aee.2018.12

Curry, G. N. (2003). Moving beyond postdevelopment: Facilitating Indigenous alternatives for “development”. *Economic Geography*, *79*(4), 405–423. https://doi.org/10.1111/j.1944-8287.2003.tb00221.x

Ekins, P. (2000). *Economic growth and environmental sustainability: The prospects for green growth.* Routledge.

Goodall, C. (2012). *Sustainability: All that matters.* Hodder & Stoughton.

Goodland, R. (1995). The concept of environmental sustainability. *Annual Review of Ecology and Systematics*, *26*, 1–24. https://www.annualreviews.org/doi/abs/10.1146/annurev.es.26.110195.000245

IPCC. (2022). *Climate change 2022: Impacts, adaptation and vulnerability. Contribution of working group II to the Sixth Assessment Report of the Intergovernmental Panel on Climate Change* [H.-O. Pörtner, D. C. Roberts, M. Tignor, E. S. Poloczanska, K. Mintenbeck, A. Alegría, M. Craig, S. Langsdorf, S. Löschke, V. Möller, A. Okem, B. Rama (Eds.)]. Cambridge University Press. https://report.ipcc.ch/ar6/wg2/IPCC_AR6_WGII_FullReport.pdf

Itkonen, P. (2022). Environmental sustainability generated by the views of the Skolt Sami and Gregory Bateson. *Journal of Ethnology and Folkloristics*, *16*(2), 290–307. https://doi.org/10.2478/jef-2022-0023

Lawson, V. (2009). Instead of radical geography, how about caring geography? *Antipode*, *41*(1), 210–213. https://doi.org/10.1111/j.1467-8330.2008.00665.x

Li, X.-F., Wang, Z.-G., Bao, X.-G., Sun, J-H., Yang, S.-C., Wang, P., Wang, C.-B., Wu, J.-P., Liu, X.-R., Tian, X.-L., Wang, Y., Li, J.-P., Wang, Y., Xia, H.-Y., Mei, P.-P., Wang, X.-F., Zhao, J.-H., Yu, R.-P., Zhang, W.-P. …, & Li, L. (2021). Long-term increased grain yield and soil fertility from intercropping. *Nature Sustainability*, *4*, 943–950. https://doi.org/10.1038/s41893-021-00767-7

Loveridge, R., Sallu, S. M., Pasha, I. J., & Marshall, A. R. (2020). Measuring human wellbeing: A protocol for selecting local indicators. *Environmental Science and Policy*, *114*, 461–469. https://doi.org/10.1016/j.envsci.2020.09.002

Mariani, M., Connor, S. E., Theuerkauf, M., Herbert, A., Kuneš, P., Bowman, D., Fletcher, M.-S., Head, L., Kershaw, A. P., Haberle, S. G., Stevenson, J., Adeleye, M., Cadd, H., Hopf, F., & Briles, C. (2022). Disruption of cultural burning promotes shrub encroachment and unprecedented wildfires. *Frontiers in Ecology and Environment*, *20*(5), 292–300. https://doi.org/10.1002/fee.2395

Maude, A. (2014b). Sustainability in the Australian curriculum: geography. *Geographical Education*, *27*, 19–27. https://agta.au/files/Geographical%20Education/2014/Geographical%20Education%20Vol%2027,%202014%20-%20Alaric%20Maude.pdf

Nieuwland, S., & van Melik, R. (2020). Regulating Airbnb: How cities deal with perceived negative externalities of short-term rentals. *Current Issues in Tourism*, *23*(7), 811–825. https://doi.org/10.1080/13683500.2018.1504899

Olson, E., & Sayer, A. (2009). Radical geography and its critical standpoints: Embracing the normative. *Antipode*, *41*(1), 180–198. https://doi.org/10.1111/j.1467-8330.2008.00661.x

Pawson, E. (2015). What sort of geographical education for the Anthropocene? *Geographical Research*, *53*(3), 306–312. https://doi.org/10.1111/1745-5871.12122

Rose, D. B. (1996). *Nourishing terrains: Aboriginal views of landscape and wilderness.* Australian Heritage Commission.

Sale, P. F. (2012). *Our dying planet: An ecologist's view of the crisis we face.* University of California Press.

Skidelsky, R., & Skidelsky, E. (2013). *How much is enough? Money and the good life.* Penguin Books.

UN World Commission on Environment and Development and Commission for the Future (UN WCED/CF). (1990). *Our common future* (Australian ed.). Oxford University Press.

Zipin, L., Fataar, A., & Brennan, M. (2015). Can social realism do social justice? Debating the warrants for curriculum knowledge selection. *Education as Change*, *19*(2), 9–36. https://doi.org/10.1080/16823206.2015.1085610

Chapter 9

Teaching for conceptual understanding

The previous chapters have examined geography's core concepts and the ways of thinking, questioning, analysing and explaining that are derived from them. However, students will need help to understand them and the ways they are used in geographical thinking. This chapter discusses some teaching strategies to achieve this aim, particularly in relation to the key concepts of place, space, environment and interconnection.

Unpacking and expressing the concepts[1]

Chapter 2 explained that these key concepts are high-level and very abstract ideas that are unlikely to mean much to students, and should not be taught on their own. Instead, Eleanor Rawling advises that:

> The key to using big concepts in a teaching and learning situation is first to build a thorough understanding of the simpler ideas in a variety of contexts. To understand space, for example, it is useful to have first understood ideas about location, distribution, pattern, interaction, distance and scale and to have studied these ideas in the context of a variety of physical and human features.
>
> (Rawling, 2007, p. 24)

Similarly, Margaret Roberts writes:

> During their practice of geography, students will gradually develop understanding of its key concepts of place, space, environment and interconnection and its many substantive concepts e.g. erosion, ecosystems, globalisation, and urbanisation. It is through repeated encounters with key and substantive concepts, applied at a range of scales in different local, national and global contexts, that students deepen their conceptual understanding.
>
> (Roberts, 2023, p. 75)

DOI: 10.4324/9781003376668-10

In Chapter 2, the key concepts were described as meta-concepts, or concepts about concepts, and needing to be unpacked for students to understand them. For English-language geography, this conceptual unpacking was begun by Eleanor Rawling (2007) and David Lambert, and later extended by them in a short list of 'major overarching generalisations that school geography can begin to build and develop with children and young people' (Lambert, 2011, p. 262). Similar generalisations were being developed for the Australian Geography curriculum at the same time, with some cross-fertilisation occurring. These are similar to 'big ideas' in science education, which are conceptual understandings that integrate ideas and describe relationships derived from a wide variety of empirical case studies[2]. Consequently they are expressed as sentences with verbs and not as topic headings or single words. Below is a set of statements that describe the main ideas within the meta-concept of place that were discussed in Chapter 4.

1 Places are parts of the Earth's surface that have been identified and given meaning by people, but these identities and meanings may differ between cultural and social groups.
2 Each place is unique in its characteristics and relationships with other places, and consequently the outcomes of similar environmental and socioeconomic processes may vary between places, and similar problems may require different strategies in different places.
3 Places provide people with the services and facilities needed to support and enhance their lives, but unequally between places and between people within places.
4 The characteristics and location of a place have an influence on the health, educational attainment, aspirations and economic opportunities of its population.
5 For many people, attachment to a place or places is important for their identity and sense of belonging, but increasing mobility and the use of telecommunication technologies may be expanding the number of places to which people feel an attachment.
6 Places can be used as laboratories for the analysis of the interrelationships between environmental and human variables, and causal relationships can be investigated through a controlled comparison of places.
7 Place provides a conceptual framework for a range of social, economic and environmental initiatives.

These statements describe the various ways in which places, as the geographical context in which we live our lives and events happen, influence our lives and these events, and they are expressions of ways of understanding the concept of place. Note that this is an example of how a key geographical concept could be unpacked and not necessarily how it should be. There is no definitive or correct way to unpack the concepts, and teachers can develop ones that they think are most appropriate for their situation.

Conceptual hierarchies

Geography's key concepts sit at the top of a hierarchy of smaller concepts, and cannot be understood without knowing the concepts on which they are built. The subject's simplest concepts are substantive or descriptive ones that help students to make sense of a collection of facts by integrating them into a single idea. The concept of weather, for example, integrates the concepts of temperature, rainfall, wind and sunshine into a single idea. Table 9.1 illustrates how this basic concept (Step 1 in the table) can be progressively combined over the school years with others to produce bigger and more abstract generalisations. Step 2 adds the concept of seasons to describe variations in weather through a year and leads to the concept of climate in Step 3. The application of the concept of space in Step 4 identifies a geographical way of explaining the characteristics of different world climatic types. The progression then moves on to the effects of climate on the environment and, more contentiously, on people and societies, and to the reciprocal effects of people on the environment. The understanding at Step 7, when combined with other progressions from the concept of environment, leads to a broad generalisation at Step 8 about human actions changing the biophysical environment, which is one of the three components of the key concept of environment described at Step 9. The other components of this concept will be contributed by progressions that start from different elements of the environment and from studies of agriculture and human societies.

Trevor Bennetts provides a good overview of the conceptual hierarchy outlined above:

> Concepts are the basic building blocks of our conceptual structures, and are usually represented by nouns and verbs. While many of the concepts which we use in geography relate fairly closely to familiar features and experiences, other concepts, which have been constructed to provide deeper insights, involve a higher order of abstraction. Thus, ideas such as suburb, town centre and commuting can be contrasted with more abstract notions, such as accessibility, urban structure and spatial interaction. Generalisations specify relationships between concepts, and are characteristically expressed by propositional statements in the form of a principle or rule. To understand the structure of an idea in the form of a generalisation, it is necessary to know the meaning of the concepts which are included in the statement and the nature of the relationships between them. Concepts and generalisations can be combined to create more elaborate conceptual structures, variously described as constructs, models and theories. While, in practice, all three terms are used in various ways, they all embody the notion of coherent conceptual frameworks, composed of interrelated ideas, which help us to organise our thinking. Therefore, they provide potentially valuable educational tools.
>
> (Bennetts, 2005, p. 115)

Table 9.1 An example of conceptual progression

Step	*Generalisation*
9	The biophysical environment supports human life and wellbeing, affects human activities and is being changed by human actions
8	Human actions are changing the biophysical environment, in both positive and negative ways
7	Human actions are changing the global climate, but differently in different places
6	The effects of climate on people and societies continues to be debated and contested
5	Climate is a major, but not the only, influence on the vegetation, soils, water resources and agriculture of places
4	World climatic types have a spatial pattern that suggests ways of explaining their characteristics
3	Climate is the average weather, with its seasonal variations, experienced by a place over a long period of time
2	Seasons describe the average weather for different periods of the year, but different cultures describe the seasons differently, using different criteria
1	Weather can be described by temperature, rainfall, wind and sunshine

Conceptual progressions such as these have an important educational role in that they lead students into more complex thinking, because each step in the hierarchy requires a synthesis of increasingly abstract ideas (Bennetts, 2005). Students start by synthesising factual information into a substantive concept, such as climate, which they then combine with other concepts, such as vegetation, to create larger and more abstract generalisations about relationships. Teachers are encouraged to construct their own progressions for the concepts examined in this book, as there are no correct or prescribed ones for any of them.[3]

Generalisations

Generalisations have been described as 'a synthesis of factual information that states a relationship between two or more concepts' (McKinney and Edgington, 1997, pp. 78–79). They are educationally valuable because they are a way of expressing students' understanding of some aspect of a core concept, as described earlier in this chapter, or of some topic they are studying, as explained below. Some generalisations are purely descriptive, such as the statement:

> Water is a difficult resource for societies to manage.

More powerful generalisations include explanations, such as the statement:

> Water is a difficult resource for societies to manage because it is variable in space and time, is integrated into and flows though the environment in complex ways, has competing uses, and is easily degraded.

The second generalisation adds the main reasons why water is a complex resource and includes many more concepts than the first one. It is a more valuable generalisation because it also can be applied to understand another context, as it tells students the possible causes and outcomes to look for.

Generalisations can be used in several ways to develop the conceptual understanding of students.

To plan a unit of work

Generalisations can be used by teachers to describe the understandings that they want their students to have at the end of a unit of work, which can then be used to plan the material that they need to study to gain them. These understandings are best developed by generalising upwards from case studies, rather than by unpacking downwards from the top with no factual knowledge of real-world examples. As Lynn Erickson, Lois Lamming and Rachel French (2017, p. 2) warn, 'One cannot understand at the conceptual level without knowing the supporting facts and skills'. Young agrees: 'Content, therefore, is important, not as facts to be memorised, as in the old curriculum, but because without it students cannot acquire concepts and, therefore, will not develop their understanding and progress in their learning' (Young, 2010, p. 25).

To focus a topic

A similar teaching strategy is to organise a topic around a generalisation, as explained in this story of a high school geology teacher:

> Having watched a river erode one part of a bank and deposit sediment a little further on, the teacher constructed the big idea that '*there are geological forces that destroy physical features and forces that create them*'. The teacher then used this idea to link and give real world relevance to a wide range of geological content that was presented in very isolated and disengaging ways in the students' textbook. For the students, framing the big idea in this way allowed them to look at everyday phenomena in new ways, attending to things they previously may not have focused upon in their environment.
>
> (Mitchell et al., 2017, p. 598)

A more geographical generalisation might express the same ideas in this way:

> Weathering and erosion lower the land surface, while the transport and sedimentation of the eroded material builds up the land surface, resulting in reduced relative relief unless interrupted by tectonic movements.

This generalisation applies the key concept of interconnection to link several geomorphic processes together to explain their combined effect on flattening the land surface, and it can be used at scales from the local to continental. It is a generalisation that applies well to the subdued land surface of the continent of Australia, where the processes it describes have been little interrupted by tectonic movements for a very long time. It could be used in teaching to organise a study that links geomorphic processes together rather than learning about them separately, as they are mostly presented in textbooks. In this approach, students may be given the generalisation at the beginning of the unit of study, rather than creating it at the end.

To help students to create their own generalisations

Another strategy is to guide students into creating their own generalisations from the content they are studying by synthesising factual details into a single sentence. The following example was mentioned in Chapter 2.

> Because of the advantages of proximity, economic activities tend to cluster in space, at all scales from the local to the global, unless tied to the location of natural resources or dispersed customers.

This is a generalisation based on the concept of space that students might create from a study of the location of economic activities, in which they discover that activities such as mining and agriculture are spatially dispersed because they depend on the location of minerals and arable land, and activities such as retailing are also dispersed because they depend on the location of customers. Many other activities, however, are not locationally restrained in these ways and instead benefit from clustering with other services, and with customers such as financial institutions and head offices who are also locationally concentrated. The generalisation enables students to reduce their study of separate economic activities into a single statement which combines spatial patterns with explanations. It could also be used to stimulate them to think of exceptions to the generalisation and how these can be explained. For example, an industry in their area may not fit the generalisation, and studying it may lead to a much more nuanced understanding of the location of economic activities, such as the role of history and individuals.

In the example above, a teacher could assist students to generalise by suggesting that they frame their study of the location of economic activities around the contrasting concepts of dispersion and agglomeration (or clustering). This would involve some initial teaching of what these concepts mean, and then students could apply them to specific economic activities. Dispersion and agglomeration are second-level concepts under the key concept of space. For other topics, teachers could suggest using one or more of the key concepts as the conceptual lens. For example, a study of global cities, such as the one described in Chapter 4 which compared the effects of globalisation on Sydney and Los Angeles, could be framed around the key concepts of interconnection and place. Erickson, Lanning and French strongly advocate this way of helping students to inductively create generalisations, when they argue that:

> When teachers share a generalization with students at the onset of the lesson and then ask students to make connections with factual examples, students are being cheated of the opportunity to construct and express the understanding

> themselves. With the appropriate instruction, students of all ages are capable of realizing the target generalizations. Students develop a deeper understanding when they are given the opportunity to do cognitive work themselves, moving from the facts and skills to transferable understandings.
>
> (Erickson et al., 2017, p. 86)

To help students to apply generalisations

The conceptual understanding of students can be deepened if they apply a generalisation to make sense of a new context or to forecast future trends. An example is the following statement that students might create from a study of coastal processes in a particular locality:

> Coastal areas are dominated by wave and tidal processes that drive weathering and sediment movement, and stopping natural sediment movements in one place may cause erosion and sedimentation in other places.
>
> (Adapted from Holden, 2011, p. 119)

This is a statement of the relationships between processes and phenomena and not a description of facts. It requires an understanding of wave and tidal processes, weathering and sediment movement, and erosion and sedimentation, which are all concepts, and it involves the key concepts of place and interconnection in that what happens in one place has effects on other places. It could be applied to forecast the effects of a proposed marina, by focusing on existing sediment movements and how they might be interrupted.

Mitchell et al. (2017) argue that effective generalisations, or 'big ideas', should meet the following criteria:

- They are central to the discipline.
- They contain many linkages or interconnections with other content.
- They are accessible to students who can relate them to their own knowledge and experience, and by doing so extend their knowledge.
- They have an emphasis on explanation.

They go on to record the things they found that teachers were doing with big ideas in concept-led teaching. Some of these are listed in the left-hand column of Table 9.2, with examples in the right-hand column that are relevant to geography.

Some writers describe generalisations as 'enduring understandings' but Milligan and Wood, writing about social studies, argue that generalisations involving people and societies should be seen as contestable, selective of particular viewpoints and subject to change over time. They warn that:

> The risk of treating concepts and conceptual understandings as [incontrovertible] is that they simply become 'facts' by another name; a conceptual approach becomes synonymous with teaching for factual understanding. In

our view, teaching concepts and conceptual understandings as static 'facts' or end-points misses the whole purpose of social studies learning. In addition, a prescriptive check-list of conceptual understandings (that learners must arrive at) risks teachers and learners overlooking rich conceptual understandings that might emerge from the learning. Without teachers understanding this, students miss out on learning for discovery and critical inquiry with an unmapped pathway ahead.

(Milligan and Wood, 2010, p. 496)

Table 9.2 Use of big ideas

Classroom use of big ideas	*Examples and comments*
Provide a need to know/source of engagement	Big ideas could be constructed to engage student interest. For example, the statement that *natural hazards are not natural* could be used to get interest in a topic on natural hazards, as it seems to be contradictory
Provide reasons for studying the domain	Asking questions that involve people, such as: • How does where people live influence their lives? • How does the biophysical environment influence people's wellbeing?
Design activities and assessment targeted to big idea	This might lead to some activities and content being dropped and others reworked. Milligan and Wood write that 'we have observed teachers reduce coverage, teach in far greater depth, and support repeated engagement with key conceptual understandings as a result of concept-led planning' (Milligan and Wood, 2010, p. 490)
Provide reasons for tools of the domain and hence equip students to use them better	A topic that requires a study of spatial distributions provides the rationale for teaching students how to map them
Link different activities. Students commonly see each activity as a discrete, isolated event, particularly if they are different types of activities	The generalisation earlier about the processes that reduce or build up the land surface links separate topics in geomorphology
Students use big ideas to frame questions that will extend/direct what is done	The generalisation earlier about the location of economic activities could generate questions about whether activities in the local area follow this pattern, or are different, and if so, why?
Asking students to bring in relevant real-life examples or applications of a big idea	Students could bring examples of economic activities they know about to discuss whether these agree with the generalisation

Source: Mitchell et al. (2017, p. 603) (left-hand column), and author (right-hand column).

Consequently they suggest teaching them as ideas for 'further inquiry and critique' rather than as endpoints.

Structuring teaching as an inquiry

Organising teaching as a guided inquiry in which students progress towards an answer to a question is another way of developing conceptual understanding and geographical thinking. This method is illustrated below, using a topic on world food security as an example. The overall question guiding the topic is:

Can the future population of the world be fed sustainably?

An answer can be developed in steps, each of them starting with a question based on one or more of the core concepts. At the end of a step, students can create generalisations that express their answer to this question, also embodying one or more of the core concepts. The steps, with their questions and possible generalisations, are described below.

Step 1. What is the source of our food?

Our food comes from the environment, either directly from plants or indirectly when we eat animal products. At a global scale, the production of plant matter, or biomass, depends on precipitation and temperature. It varies between biomes and is measured by net primary productivity. To keep the topic manageable, marine food resources are not considered. Possible generalisations are:

> The biophysical environment supports human life through the production of plant matter that supplies food, fibre and industrial materials.
> The production of plant matter is determined by temperature and precipitation, and varies spatially between biomes.

The first generalisation belongs to the key concept of environment, because it is about how the environment supports human life, while the second adds the concept of space when it mentions variations between biomes.

Step 2. What factors determine food crop yields?

Food crops are a specialised form of biomass, grown by human intervention in the environment. So the next step is to examine the environmental, economic and technological factors that influence food crop yields at a regional scale, such as irrigation, accessibility, labour supply, landforms and agricultural technologies. A possible generalisation is:

> Food crop yields are influenced at a local or regional scale by landforms, soil quality, irrigation infrastructure, labour supply, accessibility and agricultural technology, and at a global scale by temperature and precipitation.

This generalisation adds the concepts of scale and space (i.e. accessibility is a spatial concept).

Step 3. Are the methods used to produce food environmentally sustainable?

The production of food requires alterations to the environment, such as the removal of native vegetation, ploughing, terracing, irrigation, the application of fertilisers and chemicals and grazing by introduced animals. Some of these have had negative effects on the productivity of the environment, and may be a threat to the sustainability of food production. A possible generalisation is:

> Some agricultural methods have degraded large areas of the world's soils though erosion, acidification, contamination, salinisation, compaction and loss of organic matter. By reducing the productive capacity of the environment these methods are not environmentally sustainable.

Step 4. Can the world produce enough food to sustainably feed the projected future global population?

The answer to this question builds on material examined in questions 2 and 3. Students must first choose an estimate of the size of the future population of the world, about which there will be some debate, and then assess the implications of trends in food consumption, such as the growing preference for animal products. The next step is to assess whether food production can be sustainably increased to the level needed to feed that population. This increase could be achieved by expanding the cultivated area (i.e., extensification), or by increasing production on the land presently cultivated (i.e., intensification), or both, but is threatened by land degradation, industrial pollution, water scarcity, competing land uses (such as urbanisation) and climate change. A possible generalisation is:

> Intensification may be a more environmentally sustainable way to increase world food production than extensification, but the capacity of either to produce food in the future faces challenges from land degradation, water scarcity, competing land uses and climate change.

Step 5. Will increased food production provide food security for all people?

This part of the topic provides opportunities for students to evaluate and debate alternative answers, because they have to consider whether the food security of some people will be threatened more by regional factors such as poverty, conflict, water scarcity, climate change and unsustainable agricultural methods than by insufficient global food production. A possible generalisation is:

> Future world food security is threatened more by poverty, conflict, water scarcity, climate change and unsustainable agricultural methods than by the growth of population.

This generalisation uses the concepts of place, scale and human wellbeing, because it recognises that while at a global scale there may be sufficient food, at regional scales people in places with low incomes and/or conflict will be unable to obtain it.

The approach described above differs from analyses of food security that simply compare projections of future global population with estimates of potential world food output and come to the conclusion that there will be enough food. By adding ideas from the concepts of environment, sustainability, scale, place and human wellbeing, and by recognising the interconnections between them, the geographical analysis is more nuanced, and likely to be more realistic.

Structuring the teaching of the topic in this way has several educational benefits. One is that it requires students to think deeply to answer the question. They will have to understand causal relationships (such as between agricultural methods and land degradation), follow interconnections (such as between food security and conflict), evaluate opposing opinions (such as on future population growth), and recognise that food security issues vary from place to place. A second benefit is that it provides a focused guide to what students need to understand. Textbooks generally have a series of chapters on the content described above, but these are not well linked together and contain material that may not be relevant to the overall question. For example, to answer question 1, students do not need to know everything about biomes, but only what they are, their biomass productivity and why this varies between them. This reduces the quantity of information they need to comprehend, and because this material is needed to answer a question, and therefore has a clear purpose, it is more likely to be remembered. A third benefit is that an inquiry focus links the different sections of the unit together, which makes it more coherent and meaningful for students, and again more likely to be remembered. This structure could be thought of as a story, with a plot that connects the chapters and leads to a conclusion. Finally, all the generalisations are 'big ideas' that embody one or more of the core concepts.

Questioning

Asking productive questions is essential in concept-based teaching. Erickson, Lanning and French (2017) divide questions into factual, conceptual and debatable. Factual questions are needed because conceptual understanding depends on factual knowledge, but conceptual questions are essential for getting students to think conceptually. For example, the question 'what are greenhouse gases?' is a factual one. On the other hand, the question 'how does an increase in greenhouse gas emissions raise global temperatures?' is a conceptual one, because it involves the concepts of causation and interconnection. Debatable questions are also essential, because they make students review the evidence for competing positions and the strength of competing explanations. Examples might be 'does urbanisation reduce per capita greenhouse gas emissions?', or 'does food security depend more on politics than on agricultural production?'

Distinctively geographical questions are derived from the subject's concepts, such as place and space, and can be illustrated by those that a student might ask to understand why their place is like it is

- How are the characteristics of this place influenced by its relative location (concept of space)?
- How are the characteristics of this place influenced by its relationships with other places (concept of interconnection)?
- How are the characteristics of this place influenced by its biophysical environment (concept of environment)?
- How are the decisions and actions of people and organisations changing this place (concept of place)?

Conclusion

The important point in this outline of different ways of developing the conceptual understanding of students is that teaching should be aimed less at memorising the content of a topic and more at teaching students how to use geography's core concepts to ask questions, find answers, make and apply conceptual generalisations and evaluate their findings. Concept-based teaching is also claimed to improve student learning because:

- Processing factual information into concepts gives it greater meaning and makes it easier to retain.
- Perceiving patterns in what they are studying makes it easier for students to process new information.
- Synthesising facts and concepts into generalisations makes it more likely that they will be retained and used.

Summary

- Geography's key concepts of interconnection, place, space and environment are at the top of a hierarchy of subsidiary concepts and need to be unpacked for students to understand and use them.
- These subsidiary concepts can be expressed as generalisations.
- Student conceptual understanding can be developed by using generalisations to plan a unit of work or focus a topic, by helping them to create their own generalisations to express their understanding and by then applying these to new situations.
- Conceptual understanding can also be developed by organising the teaching of a topic as a guided inquiry.
- Asking conceptual and debatable questions is essential in teaching conceptual understanding.

Notes

1 This section uses material originally published in Maude (2021).
2 See Harlen (2010); Mitchell et al. (2017).
3 See Larsen and Harrington (2018).

Useful reading

Erickson, H. L., Lanning, L. A., & French, R. (2017). *Concept-based curriculum and instruction for the thinking classroom* (2nd ed.). Corwin.
A comprehensive guide to concept-based teaching.
Geographical Association. (n.d.). *Using key questions for enquiry*.

Lane, R., Carter, J., & Bourke, T. (2019). Concepts, conceptualization, and conceptions in geography. *Journal of Geography*, *118*(1), 11–20. https://doi.org/10.1080/00221341.2018.1490804

Mitchell, I., Keast, S., Panizzon, D., & Mitchell, J. (2017). Using 'big ideas' to enhance teaching and student learning. *Teachers and Teaching*, *23*(5), 596–610. https://doi.org/10.1080/13540602.2016.1218328
Available from ResearchGate by searching for the title.

Roberts, M. (2017). Planning for enquiry. In M. Jones (Ed.), *The handbook of secondary geography* (pp. 48–59). Geographical Association.
A thoughtful review of the meaning of inquiry.

Marschall, C., & French, R. (2018). *Concept-based inquiry in action: Strategies to promote transferable understanding*. Corwin.
A very practical book on teaching concept-based inquiry.

References

Bennetts, T. (2005). Progression in geographical understanding. *International Research in Geographical and Environmental Education*, *14*(2), 112–132. https://doi.org/10.1080/10382040508668341

Harlen, W. (Ed.) (2010). *Principles and big ideas of science education*. Association for Science Education.

Holden, J. (2011). *Physical geography: The basics*. Routledge.

Lambert, D. (2011). Reviewing the case for geography, and the 'knowledge turn' in the English National Curriculum. *Curriculum Journal*, *22*, 243–264. https://doi.org/10.1080/09585176.2011.574991

Larsen, T., & Harrington, J. A. (2018). Developing a learning progression for place. *Journal of Geography*, *117*(3), 100–118. https://doi.org/10.1080/00221341.2017.1337212

Little, C. (2017). Designing and implementing concept-based curriculum. In L. S. Tan, L. D. Ponnusamy, & C. G. Quek (Eds.), *Curriculum for high ability learners* (pp. 43–59). Springer Nature.

Maude, A. (2021). Recontextualisation: Selecting and expressing geography's 'big ideas'. In M. Fargher, D. Mitchell, & E. Till (Eds.), *Recontextualising geography in education* (pp. 25–39). Springer.

McKinney, C., & Edgington, W. (1997). Issues related to teaching generalizations in elementary social studies. *The Social Studies*, *88*(2), 78–81. https://doi.org/10.1080/00377999709603751

Milligan, A., & Wood, B. (2010). Conceptual understandings as transition points: Making sense of a complex social world. *Journal of Curriculum Studies*, *42*(4), 487–501. https://doi.org/10.1080/00220270903494287

Rawling, E. (2007). *Planning your key stage 3 geography curriculum*. Geographical Association.

Roberts, M. (2023). Powerful pedagogies for the school geography curriculum. *International Research in Geographical and Environmental Education*, *32*(1), 69–84. https://doi.org/10.1080/10382046.2022.2146840

Young, M. (2010). The future of education in a knowledge society: The radical case for a subject-based curriculum. *Journal of the Pacific Circle Consortium for Education*, *22*, 21–32. http://programs.crdg.hawaii.edu/pcc/PAE_22__1__final_10.pdf

Chapter 10

Land cover change

Have we cleared too much of the earth's vegetation?

The topic is expressed as a question to be debated; not as content to be learned.

First step: What is land cover change, and why is the question worth examining?

What is land cover?

The question is about changes to what scientists call land cover. This is the physical and biological cover of the land surface of the Earth. It includes native vegetation, crops, planted forests, urban areas, bare land, ice, snow and water in lakes and dams. Thousands of years ago, the Earth's land surface was mostly covered with various types of vegetation, but people have been transforming this covering for a very long time. Early societies burned the vegetation to attract game, cleared land for farming, constructed irrigation systems in arid regions and built settlements. More recently, industrialisation, economic expansion and population growth have accelerated land cover change through increased agricultural production, the growth of cities and industrial areas, mining and the construction of roads and railways (Ellis, 2021; Ellis et al., 2013). These changes to the surface of the Earth may seem unremarkable, because they have produced the environment with which we are familiar, but they are starting to have some major consequences, as this quotation by a group of geographers explains:

> Changes in the terrestrial surface of the Earth characterize the history of humankind. Today, the Earth's landscapes have been largely transformed, restructuring ecosystems and their services in the process of supporting a population approaching eight billion at the highest average level of material consumption in human history. The aggregation of these changes creates environmental impacts that challenge the fundamental structure and function of the earth system, with cascading effects on biodiversity, biogeochemical cycling, and climate change.
>
> (Turner II et al., 2020, p. 489)

DOI: 10.4324/9781003376668-11

The quotation describes land cover change as now being extensive enough to have significant global as well as local environmental consequences, among them being its contribution to climate change, so the question is clearly a significant one worth examining. As explained in Chapter 6, interest in land cover change is not new. *Man's role in changing the face of the earth*, the record of an international symposium, was published in 1957, and over the last three decades, a growing and interdisciplinary field of land cover/land use science has developed. Geographers have contributed to this development, but in schools, study of the topic is fragmented between several sections of the curriculum, and the concept of land cover is barely mentioned in textbooks. This is a loss, because land cover provides a good opportunity to use geographical thinking to understand a major current issue of both local and global significance.

This chapter focuses on rural areas, as the situation in urban areas has already been discussed in Chapter 6. It begins with an overview of the extent of change to the surface of the Earth. The main change has been a major decline in the area of forest and woodland, and this has had an impact on weather and climate, atmospheric CO_2, water resources and water quality, biodiversity and other environmental characteristics. The chapter then applies the concepts of sustainability and human wellbeing to argue that people *have* cleared too much of the earth's vegetative cover, and that this needs to be reversed. The reasons why people clear land are then examined, as strategies to prevent further clearing must address its causes if they are to be effective. The chapter concludes with a discussion of the various strategies that have been developed to reverse vegetation loss, their objectives and limitations and the different viewpoints on which ones should have priority.

Next step: The extent of land cover change must be understood before students can examine its effects.

The extent of land cover change

Figure 10.1 shows estimates of changes to the land cover of the Earth over the past 10,000 years. The estimates are for habitable land areas, and exclude deserts, glaciers, rocky terrain and other barren land, which needs to be remembered when comparing these estimates with others that do not. The figure shows that the area of forest has been reduced from 57% of the habitable land to 38%, and that almost half of the habitable surface of the Earth has been converted to cropland and grazing land.

At a national or regional level, the extent of change may be much greater. In the agricultural areas of the southwest and southeast of the Australian continent, for example, approximately 50% of native forest and 65% of native woodland has been cleared or severely modified. Furthermore, in many areas where the type of land cover has not changed, it has been degraded by

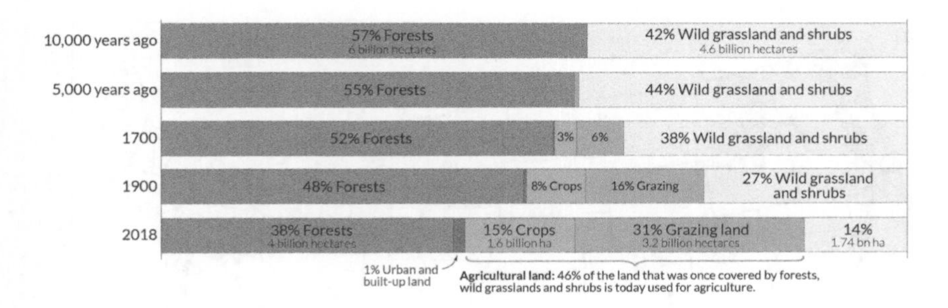

Figure 10.1 Historical change in world land cover.

Source: Ritchie and Roser (2021a).

Reproduced under Creative Commons.

the way the land is used, such as from the expansion of grazing into areas of grassland and shrubland. Similarly, in the conterminous USA an estimated 53% of the combined area of forest, grassland and shrubland was converted to cropland, pasture or urban development between 1630 and 2020 (Li et al., 2022). Figures 10.2a and 10.2b graphically portray the extent of this land cover change by comparing the forest area of the USA in 1620 and 1920.

Next step: The question asks whether we have cleared too much of the Earth's vegetation. To determine whether this is the case, the effects of land cover change must be identified and evaluated.

What are the effects of land cover change on the environment?

The changes in land cover described above affect the environment in a number of ways. These can be understood by using the concept of interconnection, because changing one component of the environment will change other components with which it is interconnected through cause-and-effect relationships.

Weather and climate

Most of the focus in land cover change research has been on the effects on weather and climate, as the removal of trees and shrubs and their replacement by crops and pastures can change the flows of energy and moisture between the land surface and the atmosphere, and in turn produce changes in weather and climate. This is the result of three changes to the land surface: evapotranspiration, albedo and surface roughness.

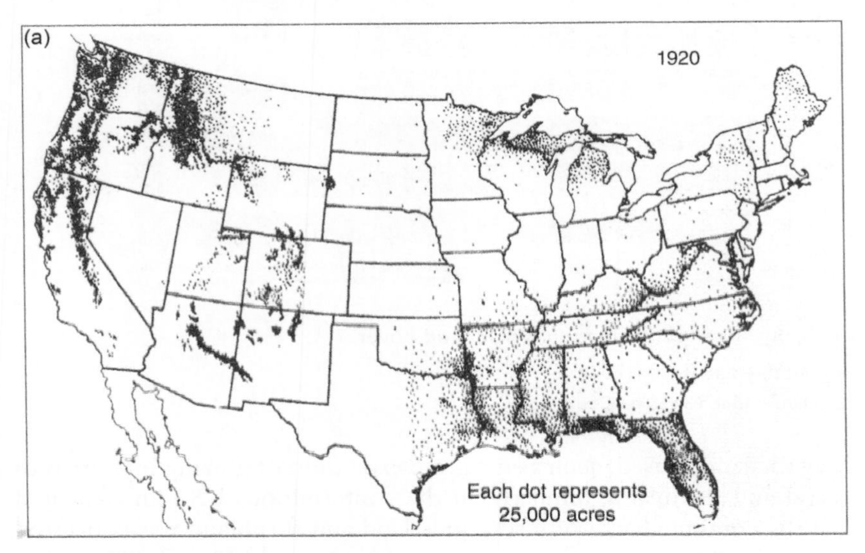

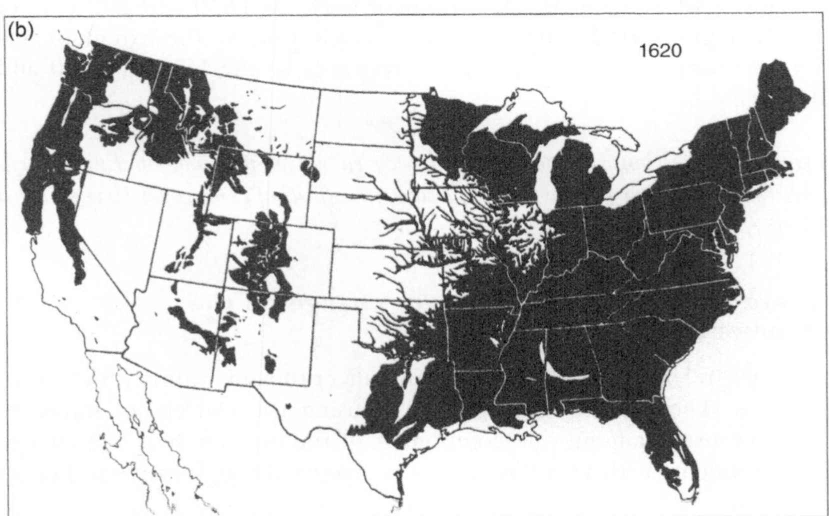

Figure 10.2 The spatial distribution of natural forest in the USA in (a) 1920 and (b) 1620.

Source: Reproduced from Goudie (2019, p. 41) with permission of the Licensor through PLSclear.

Evapotranspiration from native vegetation produces a flow of moisture from the land to the atmosphere, and if the vegetation is cleared, this flow will be reduced. The extent of this reduction varies spatially with latitude. Reductions are greatest in the tropics and least in high latitudes, with temperate regions in between, because rates of evapotranspiration from the native vegetation tend to decrease as latitude increases. Evapotranspiration also cools the land surface, so if it is reduced by land cover change surface temperatures will rise. In general, then, if crops and pastures have lower evapotranspiration rates than the native vegetation they replace, land cover change will reduce evaporative cooling and the transpiration of moisture and produce a hotter and drier land surface and atmosphere. On the other hand, if irrigated crops replace the native vegetation in low rainfall areas, evapotranspiration will be increased and precipitation may increase downwind. This outcome has been identified in the Ogallala aquifer region of the USA, where irrigation has resulted in a 15% increase in July precipitation for several hundreds of kilometres downwind of the region. Large dams have also been shown to increase evaporation and downwind rainfall.

The effects of reduced evapotranspiration are not only local. Globally it is estimated that evapotranspiration from land surfaces contributes the water vapour for 40% of the world's rainfall over land, with the remaining 60% coming from water vapour transported from the seas and oceans, so deforestation has significant consequences for rainfall and water resources in other areas.

Albedo is a measure of the solar radiation reflected back into the atmosphere by the land surface, and consequently of the amount of unreflected radiation left to be absorbed by and warm the land surface. Snow, for example, has a high albedo and reflects most of the solar energy falling on it, while at the other extreme dark road surfaces have a low albedo and quickly warm up in the sun. A consequent generalisation is that if the cropland or pasture absorbs less radiant solar energy than the native vegetation it has replaced, because it is more reflective, surface temperatures will be reduced. As a result, the convective upward movement of warmed air into the atmosphere will be reduced, which will decrease the chance of rainfall.

The roughness of the land surface has an effect on the vertical movement of air from the surface into the atmosphere. Rough surfaces, such as forests, interrupt the smooth flow of air and create turbulence and rising air, while smoother surfaces, such as annual crops and pastures, produce less turbulence. If land cover change leads to reduced turbulence, the transport of water vapour into the lower atmosphere will decrease and rainfall will be reduced.

The combined effect of the three processes described above on temperature varies with latitude, because a reduction in evapotranspiration increases temperatures while an increase in albedo reduces them. Simulations of the effects of extensive deforestation estimate that in tropical areas the effects of lost roughness and lost evapotranspiration are stronger than the effects of changed albedo, resulting in surface warming. In high latitude regions, on the other hand, the effects of albedo dominate, and the net effect of deforestation is cooling. In mid-latitude regions, the net effect varies above and below zero.

The outcomes of the processes on rainfall are more complex. Temperate southwest Australia provides an example, where research has estimated that somewhat over half of the 15–25% decline in rainfall in the Wheatbelt Region has been caused by the removal of native vegetation for grain farming. This decline has been produced by three processes. First, it is estimated that the replacement of the native vegetation with annual crops and pastures over the past 200 years has reduced the flow of evapotranspiration from the land to the lower atmosphere by about 10%. Second, the farmed land has a higher albedo than the native vegetation, absorbs less solar energy and consequently produces less rising air through convection. Third, crops and pasture have a smoother surface than the native vegetation and create less turbulence, which further reduces the upward movement of air. Rising air carries moisture to heights where convective clouds will form, and some may produce rain, so if evapotranspiration is reduced and there is less upward movement of air carrying this moisture, rainfall will be correspondingly reduced. Figure 10.3 illustrates this, as the satellite image shows cloudless skies over farmland, and clouds over the uncleared land inland of the rabbit proof fence that marks the boundary of cultivation, but not necessarily of rabbits.

For the Australian continent as a whole, researchers have estimated the change in continental water vapour flow (evapotranspiration) over the past 200 years. They write:

> During this period there has been a substantial decrease in woody vegetation and a corresponding increase in croplands and grasslands. The shift in land use has caused a ca. 10% decrease in water vapour flows from the continent. This reduction corresponds to an annual freshwater flow of almost 340 km^3. The society-induced alteration of freshwater flows is estimated at more than 15 times the volume of run-off freshwater that is diverted and actively managed in the Australian society.
>
> (Gordon et al., 2003, p. 1973)

Another example comes from Europe. Researchers compared several thousand pairs of sites across the continent, with each pair containing a forested site and an unirrigated agricultural site that was otherwise geographically similar. Rainfall was observed to be consistently higher over the forested sites, and the researchers went on to suggest that the summer droughts predicted to increase with global warming could be mitigated by tree planting (Baker, 2021).

Finally, in the tropical Amazon Basin, where it is estimated that from 50% to 70% of rainfall comes from the moisture produced by evapotranspiration, modelling suggests that deforestation may reduce local rainfall by about 25% annually. Also, because the Amazon Basin exports about a third of the moisture it produces through the flow of air to the south, land clearing in the region may be contributing to the severity of droughts in southern Brazil. Another study estimates that extensive tropical deforestation, by reducing evapotranspiration, can decrease regional rainfall by up to 40%.

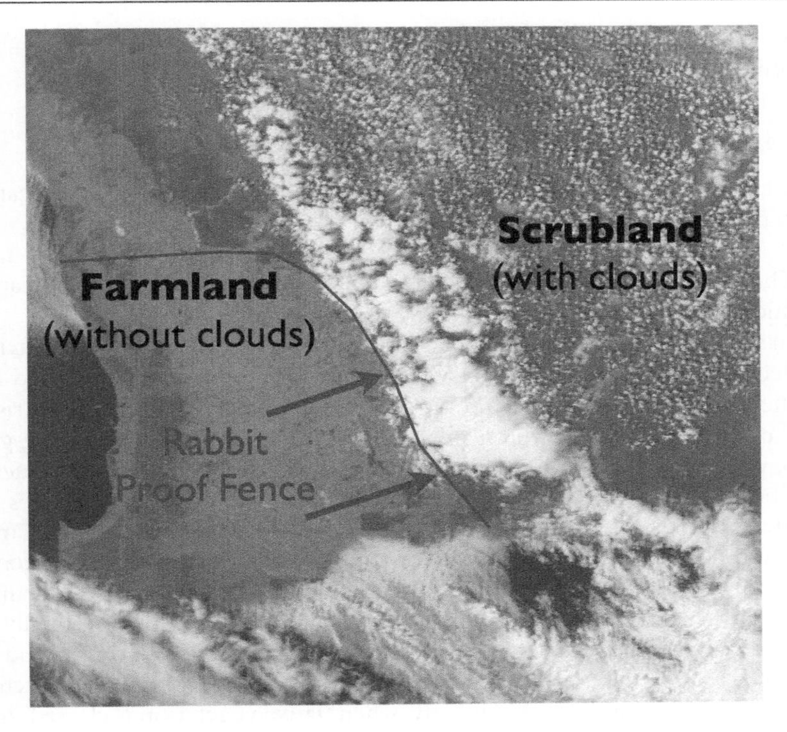

Figure 10.3 Convective clouds forming over uncleared land, Western Australia.

Source: ©UCAR Reproduced with permission. https://www.eol.ucar.edu/content/research-goals-objectives.

Carbon emissions

A second climatic consequence of land clearing is its effects on the carbon cycle. Plants store large quantities of carbon, equivalent to roughly three-quarters of the quantity stored in the atmosphere, and well over half of this is stored in forests. Plants also remove large quantities of carbon from the atmosphere each year through photosynthesis, amounting to about 20% of carbon emissions from all sources, natural and human-induced. Forests, especially those in the tropics, are the major contributors to this sequestration. Carbon is also absorbed by the oceans, and on land is returned to the soil through litterfall. Combined, these processes have kept atmospheric CO_2 levels stable for millions of years, but this stability is now threatened by two new sources of carbon emissions—fossil fuels and land cover change. When vegetation is cleared, whether for timber or farming, some of the vegetation is burned or left to decay, and this releases carbon into the atmosphere. As a result, land cover change is estimated to contribute around 30% of total human-induced carbon emissions, and a modelling study which added CO_2 emissions to the other effects of land cover change found that deforestation adds to global warming at all latitudes from 50°S to 50°N (Lawrence et al., 2022).

Other impacts of land cover change on the environment

Climate is not the only part of the environment affected by land clearing. Other impacts include:

- The clearing of forest for agriculture increases soil erosion, which can reduce future food production.
- In some parts of the world, land clearing has resulted in soil salinity, after the deep-rooted trees and shrubs, whose extraction of ground water had kept the water table well below the surface, were removed. This caused a rise in the water table, and if this water was salty because of the underlying geology, the top layer of the soil became saline and agriculturally unproductive.
- The clearing of forests and shrubland damages the ecosystem services that they provide, such as protecting watersheds, cooling environments, moderating floods, providing habitat for animals and supplying valuable genetic materials. For example, the removal of native vegetation increases runoff, which in turn increases the likelihood of flooding and its potentially destructive impacts. One study estimates that in tropical countries, flood frequency increased and floods lasted longer as natural forest cover declined between 1990 and 2000. Similarly, when native vegetation is cleared, habitat for both fauna and flora is lost and fragmented, threatening the survival of many species. Intriguingly, however, studies of sun bears in Malaysia found that they preferred regenerating forests that had been selectively logged to adjacent primary forests, because the former had richer supplies of their food. We cannot assume that environmental change is necessarily bad for wildlife.
- Land cover change is the major cause of the loss of biodiversity. A global study of the effects of land cover change found that in 28% of the grid cells in the research the resulting reduction in local species richness (the number of different species in a defined area) had exceeded 20%. The significance of this finding is that a reduction of over 20% in species richness may be sufficient to reduce biomass production and the provision of ecosystem services such as pollination.
- Land clearing changes freshwater quality, when erosion caused by the removal of vegetation adds sediments to streams, and when agricultural chemicals used in farming cleared land are washed into waterways. In the State of Queensland in Australia, for example, land clearing has greatly increased sediment and chemical runoff into the coastal waters around the Great Barrier Reef, damaging the health of the reef's ecosystems and reducing their tourism value.

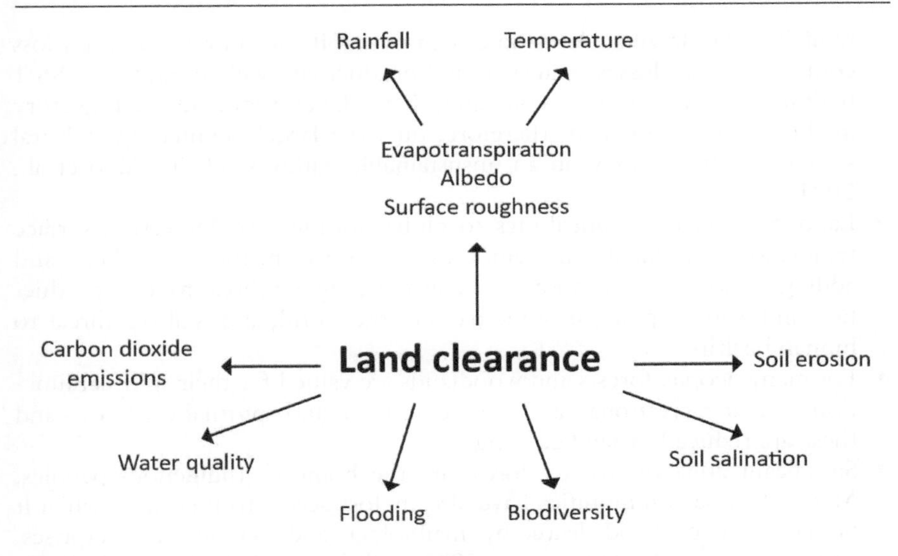

Figure 10.4 The environmental effects of land clearance.
Source: Author.

Understanding of the effects of land cover change outlined above is diagrammatically represented in Figure 10.4.

The next step is an evaluation of these environmental effects, using the concepts of sustainability and human wellbeing, in order to decide if change has been too much.

Has there been too much clearance of the Earth's vegetation?

'Too much' implies that land cover change has created significant problems, and the concepts of sustainability and human wellbeing, discussed in Chapter 8, provide ways of evaluating whether this is the case. Sustainability is about ensuring that changes to the environment do not threaten the important services that the environment provides to sustain human life and wellbeing. Land cover change affects the sustainability of these services in a number of ways:

- The reductions in rainfall and increases in temperature of the kind described above threaten the ability of the environment to produce food and supply water. For example, a study of the effects of deforestation on

rainfall in the Brazilian Amazon estimates that if current rates of forest loss continue, future losses in agricultural productivity will amount to US$ 1 billion annually. The study concludes that 'The current land-use trajectory in the Brazilian Amazon, therefore, puts the largely rainfed agricultural systems of the country on an unsustainable pathway' (Leite-Filho et al., 2021, p. 5).
- Land cover change contributes to global warming by increasing surface temperatures, reducing the removal of carbon from the atmosphere, and adding carbon to the atmosphere. This warming is a threat to food production and water supplies in some parts of the world, and is also a threat to human health.
- For many people, forests and woodlands are valued for their spiritual functions—their recreational, aesthetic, emotional and spiritual qualities—and these are reduced by land clearing.
- Significant areas of tropical forest are the home of Indigenous peoples. Many of these communities have already lost access to their land when it has been occupied and cleared by smallholders and commercial enterprises, and have been reduced to poverty. The wellbeing and even survival of others is threatened by further deforestation.

These conclusions suggest that vegetation loss now threatens the sustainability of a range of environmental services that support human life, and is a threat to the wellbeing of many people. A strong case therefore can be made for strategies to reduce further loss of native vegetation and to restore vegetation where possible. As argued in Chapter 2, the selection of these strategies depends on a clear understanding of the causes of vegetation loss, as strategies must address the causes.

Next step: Students must understand why people are clearing land before they can evaluate strategies to change their behaviour.

Explaining land cover change

People, acting as individuals, households and businesses, are the immediate agents of land clearing, but their actions are influenced by a wide range of underlying factors, from population growth to commodity prices and government policies. How can these causal factors be identified? A major contribution of geographers has been to explore a variety of potential causal relationships in a holistic way, employing a variety of methods ranging from the highly quantitative to the highly qualitative.

One approach has been to use quantitative methods that analyse spatially coded data on land cover and selected explanatory variables. These studies use information from satellite images to produce measures of land cover change for areas as small as a pixel or as large as a country, and combine them with remotely sensed data on environmental characteristics such as rainfall, slope and soils, and socioeconomic information on roads, population characteristics, poverty, agricultural prices and other characteristics. This research has been described as 'socializing the pixel' and 'pixelizing the social' (Lambin and Geist, 2006, p. 5). Spatial statistical methods are then used to identify relationships between the dependent variable (land clearing) and the independent explanatory variables. These methods can test explanations of current land clearing, and predict the location of future clearing, and in this they have been moderately successful. Studies within countries add additional factors, such as road construction, timber value, rainfall, topography, population density, income levels and the location of Indigenous lands.

A limitation of these methods is that they do not focus on the individual agents of deforestation, which are farmers and their households. A study by geographers that did analyse data on individual properties and households was an investigation of deforestation by peasant colonists on the forest frontiers of Brazil (Caldas et al., 2007). This combined survey data on the characteristics of household economies (including distance from the main highway) with measures of deforestation from remotely sensed imagery. The research concluded that the extent of deforestation was particularly related to the household labour force (the number of men) and distance from the Transamazon Highway. The second of these reflects the effects of market access, as measured by distance, on the increase in land value that can be gained by forest clearing. Land closer to markets is more profitable and so commands higher prices, so there is an incentive to clear it for sale.

Quantitative methods have some limitations. One is that the variation in the rate of deforestation 'explained' by the statistical analysis is often limited, suggesting that there are other causal factors that have not been included in the modelling. Another is that appropriate socioeconomic data are not always available, and some explanatory factors, such as culture, are difficult to quantify. A final limitation is that while numerical modelling is able to identify statistical associations between variables, it cannot identify the cause-and-effect mechanisms that create these associations, and consequently cannot explain why land change happens.

A second approach in research on the causes of vegetation clearance has been through qualitative case studies. These have been able to identify the sequence of events that produce deforestation, the motives of the actors involved and the underlying causes that are difficult to include in models because they cannot be quantified or mapped. These methods are the most useful for identifying policies to limit land clearing because they can explain why people are clearing land and therefore what measures might change their behaviour and are place-specific. As McCarthy and Tacconi argue:

> Addressing the issue of tropical deforestation with reasonable policy advice will continue to be plagued by the historical tendency to explain deforestation as being caused by universal factors, leading in turn to the adoption of universal solutions to problems that may not in fact be universally present. … Policies developed to tackle deforestation must in the end be case-specific, targeted towards local circumstances …
>
> (McCarthy and Tacconi, 2011, p. 128)

Geographers will interpret this comment as support for the importance of 'place'.

Table 10.1 simplifies and summarises the findings from both types of research. It demonstrates that there is no single cause of land clearing, with different agents responding to different causal factors, and several factors combining in each case. Research also shows that there are complex connections and interactions between the many causes of land cover change, and because these vary between places, it is difficult to generalise and predict future change. Strategies to manage land cover change have to be sensitive to the differences between places.

Some of the items in the table need a brief explanation.

Table 10.1 Explaining land clearing

<table>
<tr><th>Agents</th><th>Motivation</th><th>Underlying causal factors</th><th>Facilitating factors</th></tr>
<tr><td>Small semi-subsistence farmers</td><td>Escape poverty</td><td>Poverty
Land settlement schemes
Road construction</td><td>Unequal land ownership
Population growth
Land that could be cleared</td></tr>
<tr><td>Plantation and logging companies</td><td>Profit</td><td rowspan="2">Domestic demand from growing urban populations
Global demand
Road construction</td><td rowspan="2">Land that could be cleared
Insecure land tenure
Weak government regulation
Government economic policies
Displacement of land use Culture</td></tr>
<tr><td>Commercial farmers</td><td>Profit
Land value appreciation</td></tr>
</table>

Escape poverty

In a study of the role of peasant farmers in Amazonian deforestation, the authors conclude:

> Of particular importance are the underlying factors that bring colonists to the Amazon forests in the first place. They do not arrive by accident, and typically are responding to social and economic circumstances beyond their control, such as a life of poverty or a lack of opportunity in other parts of the country. [We are] simply describing a final result in a long chain of action and reaction, stemming from the power centers of decision making in the national economy and government.
>
> (Caldas et al., 2007, p. 103)

What the authors are pointing out is that deforestation caused by households migrating from other areas to the forest frontier has to be tackled at the source of the migration, by addressing problems such as unequal land ownership and lack of economic opportunity, and ultimately at a governmental and political level.

Land value appreciation

Several studies of regions within the Amazon Basin found that rising land values had increased the incentives for farmers to clear land which they could later sell at a higher price. The study by Caldas et al. noted earlier showed that this incentive varies with market access, as measured by distance.

Road construction

Road construction made forest areas accessible to migrants and commercial enterprises and made it economic to transport products to local, national and international markets. Roads were constructed by governments and by private logging companies who built networks of unofficial roads into the forests from the government roads.

Population growth

A number of studies have investigated whether land clearance is caused by population growth but most fail to find a positive relationship. Some point out that areas with high rates of deforestation, such as Amazonia and much of Southeast Asia, have low population densities, and that in these regions, the agents of land clearance have increasingly been commercial ranchers and plantation companies. High population densities in regions such as Java, on the other hand, have contributed to migration to forest areas in other parts of Indonesia, but this often required government assistance through road construction and settlement schemes. A recent research project concluded that their analysis shows that:

> forest loss is positively correlated with urban population growth and exports of agricultural products for this time period. Rural population growth is not associated with forest loss, indicating the importance of urban-based and international demands for agricultural products as drivers of deforestation. The strong trend in movement of people to cities in the tropics is, counterintuitively, likely to be associated with greater pressures for clearing tropical forests.
>
> (DeFries et al., 2010, p. 157)

Displacement of land use

High-income country policies to protect forests or to mandate the use of biofuels have displaced production to lower-income countries. For example, a study of small farmers in the Amazon found that they were being pushed into the forest frontier by cattle ranching taking over their land, because ranching was being displaced by more profitable soybean production, whose growth was a response to international demand created by the replacement of soybean production in the Midwest of the USA with maize production for ethanol. The latter had been stimulated by US Government policy that required the addition of small proportions of ethanol to petrol. This situation is a good example of the unintentional effects of interconnections.

Insecure land tenure

Much of the tropical forest frontier is public land, and farmers migrating into these areas and occupying land do not have formally registered ownership. This may lead them to clear land as a way to establish their claim to it, which increases the extent of deforestation.

Weak government regulation

Examples of the effects of government policies on deforestation are described later in the chapter.

Culture

The author of a landmark historical geography of world deforestation argues that 'nonmaterial forces like motivation, sentiment, symbolism (especially strong in forest environments), political, religious and social mores are largely ignored [in analyses of deforestation] because they cannot be quantified or factored into explanation easily' (Williams, 2006, p. xvi). A specific example of a cultural influence comes from research in western Brazilian Amazon which found that:

> cattle raising [a major contributor to deforestation] must be understood in relation to 'cattle culture,' or cultural beliefs and practices that promotes cattle raising as a modern and desirable way of life. Development and popular narratives promote cattle raising as the pathway to economic growth, national security, and individual socio-economic mobility. ... The preference for cattle-based livelihoods is also reflected in perceptions of the landscape: many stakeholder groups associate cattle pastures with positive social values, such as social status and progress, whereas forests are associated with poverty and decline.
>
> (le Polain de Waroux et al., 2021, p. 453)

Differences between world regions and countries

The causes of tree cover loss vary between world regions and countries. One study (Curtis et al., 2018, Table S7) used Google Earth imagery to classify tree cover loss between 2001 and 2015 by driver and concluded that:

- In Latin America, the main drivers were commodity-driven deforestation (56%), mainly for beef cattle and soybeans, followed by shifting agriculture (31%) and forestry (13%).
- In Southeast Asia, the main drivers were commodity-driven deforestation (78%), mainly for oil palms, followed by forestry (13%) and shifting agriculture (9%).
- In Africa, the main driver was shifting agriculture (92%).
- In North America, the main drivers were forestry (56%) and wildfires (40%). Wildfires are a cause of tree cover loss that has not been noted so far in this chapter, but has become more widespread in recent years.
- In Australia/Oceania, the main drivers were wildfires (53%) and forestry (29%).

A consequence is that different strategies will be needed in each region because of the different causes.

Students can now evaluate the strategies that can be used to reduce the negative effects of land cover change on the environment and its ability to support human life and wellbeing.

Strategies to reduce or prevent further vegetation clearance

In reviewing these strategies students should think about how each one addresses the causes of land cover change, as this is necessary for them to be effective.

Reduce land clearance by regulation

The availability of land that could be cleared for timber or farming is an obvious factor in land cover change, and a response has been to reduce its availability by restricting or prohibiting land clearance. Much of the remaining world forest land is publicly owned, so governments have the authority to control its use, but they can also regulate vegetation clearance on privately owned land. However, such regulation can be politically difficult. For example, in Brazil, deforestation rates declined between 2005 and 2012 under a government that created new conservation areas, strengthened law enforcement, and adopted other measures noted later. Since 2012, however, deforestation rates have increased, and a significant part of this increase has been attributed to:

> the political power of the 'ruralists,' a coalition of legislative representatives of large landowners and agribusiness interests. These politicians have taken the lead in the National Congress and since 2012 have been pushing constitutional amendments that weaken environmental protection and facilitate infrastructure development and agribusiness.
>
> (Carrero et al., 2020, p. 978)

This pro-deforestation trend was greatly strengthened by Jair Bolsonaro, who was elected President in 2019 and was largely opposed to environmental protection. The re-election of a previous President in 2022 may see a return to stronger government action.

The State of Queensland in Australia provides a similar example. Over a nearly 20-year period, the area of woody vegetation cleared in the State fell dramatically from 758,000 hectares in 1999–2000 to 78,000 hectares in 2009–2010, in response to tighter controls. A conservative government in office between 2012 and 2015 greatly weakened the legislation, and land clearance jumped to 295,000 hectares in 2013–2014. The following Labor Government has failed to adequately restore controls and close loopholes in the legislation, and land clearing in 2018–2019 is estimated to have reached 680,688 hectares (Wilderness Society, 2022). Like Brazil, Queensland has a strong farming lobby.

Another issue with the protection of forest and woodland areas is that many such areas are the home of Indigenous communities who depend on them for their livelihood. In some cases, these communities have been expelled from their land on the grounds that their continued presence was incompatible with the preservation of the area's biodiversity. Such actions are not only socially unjust, but also overlook 'the fact that many of these "high-value," biodiverse landscapes are the historical product of, and thus require, human intervention to maintain the very values for which they are lauded' (Fletcher et al., 2021). In some countries, Indigenous groups have been given legal title for the lands they occupy and assistance to protect their forests, and several studies show that rates of deforestation are lower in areas managed by Indigenous peoples than in state-managed protected areas. An ingenious strategy tested in the Peruvian Amazon provided Indigenous communities with real-time satellite data about possible illegal activities in their area, downloaded onto an app on their smartphones. The communities then sent patrols to investigate, and the result was a substantial reduction in deforestation. Many countries, however, still refuse to recognise the land rights of Indigenous peoples or to adequately support their role as forest protectors.

Reduce migration to forest areas

Migration to clear land and establish small farms is a significant cause of deforestation in many tropical regions. To reduce this demand for land, strategies to reduce poverty in the sending areas or to create alternative migration opportunities are needed. This again requires political leadership and action.

Reduce the profitability of land clearance

A major cause of land clearance is its profitability. If this can be reduced deforestation should decline, and this outcome can be achieved in several ways:

- By removing subsidies that increase the profitability of agricultural production.
- By excluding farmers who clear land from receiving agricultural credit.
- By not constructing roads that improve access to forest areas.

Reduce the need to clear more land

If farm productivity per hectare can be increased, the need to clear more land to produce more output will be reduced. This can be achieved by assisting farmers to use improved farming methods or plant different crops. This strategy is particularly relevant where loss of soil fertility or poor farming practices on previously cleared land leads farmers to clear more forest to restore levels of production.

Supply chain strategies

A strategy tried with internationally traded products such as palm oil, timber and cacao is for importing countries and companies to require certification that commodities have not been produced from newly cleared land. Companies may choose to adopt this strategy without government involvement because of demand from their customers, who want to know that the products they buy have been sustainably produced. This is where non-government organisations and households can have an influence on deforestation. However, much of the demand for products from deforested land is domestic and less likely to be subject to certification programs. For example, one study estimated that international trade was only responsible for a third of tropical deforestation, a proportion that ranged regionally from 9% in Africa to 23% in Latin America and 44% in the Asia-Pacific region. (Ritchie and Roser, 2021a)

Financial incentives to preserve forests

The financial profitability of vegetation clearance can also be countered if landowners are paid to preserve forests and woodland on their properties for the ecosystem benefits they provide and the carbon emissions avoided or if countries are rewarded for reducing deforestation. Some programs are national, with payments made by governments to eligible landowners. Others are international, such as the Reducing Emissions from Deforestation and Degradation (REDD+) program, in which payments are made by high-income countries to lower-income countries that achieve agreed performance targets for forest preservation and forest management. So far the results of these programs have been modest.

The strategies outlined above are designed to reduce further conversion of forests and woodland to cropland and pasture. Their potential is demonstrated by the marked reduction in deforestation in Brazil mentioned earlier, which has been attributed to these initiatives:

> Establishment of new protected areas (including indigenous reserves) across the path of the advancing agricultural frontier—the 'arc of deforestation'—and enhanced law enforcement, aided by remote sensing technology, slowed illegal logging and clearing for cattle pasture. Soy traders, facing potential loss of access to international markets due to advocacy campaigns, imposed a moratorium on sourcing from recently deforested land. The moratorium, augmented by a cutoff of state agricultural credit to municipalities with high deforestation rates, resulted in a decline in forest clearing.
>
> (Seymour and Harris, 2019, p. 756)

Strategies to increase the area of forest and woodland

Strategies to reduce deforestation can be complimented by ones to increase the area of forest and woodland.

Afforestation and Reforestation

Afforestation is the planting of trees on land that has not been forested for a long time or ever, while reforestation is the re-establishment of forest on land that was previously forested but has been logged, temporarily farmed or degraded. Reforestation can be achieved either by planting new trees, or by allowing cleared vegetation to regenerate naturally. These are popular strategies to address the environmental problems created by land cover change, but have some limitations and need careful planning.

One constraint is that enough cropland and pasture must be kept to provide for human needs for food and fibre. Intensification of farming on the most productive land could reduce the growth in the area of cropland required in the future, as the number of people supported by the same area of cropland has been steadily increasing. A decline in beef consumption through changed diets might reduce the area of pasture required, but there is no sign of this happening at present. There are consequently significant constraints on how much land can be forested.

A second limitation on the area of land that can be forested is that natural grasslands are generally not suited to afforestation. Veldman writes:

> Compared with grasses and forbs, trees require far more water and soil nutrients and have markedly different patterns of above- and belowground carbon allocation. Consequently, afforestation and forest expansion can dramatically alter nutrient cycles, reduce soil-carbon storage, and change hydrology.
>
> (Veldman, 2015, p. 1011)

This quotation suggests that the contribution of grassland afforestation to carbon sequestration may be limited, and in areas where fire is a risk it may also be insecure. Grassy biomes, on the other hand, can store a lot of carbon, but mostly below ground where it is secure. Trees planted in grassland areas are also a hydrological risk, as they increase the evapotranspiration of soil moisture, and cases have been reported from China where afforestation programs have resulted in water shortages and tree mortality (IPCC, 2019, p. 98).

Some scientists argue that instead of planting more trees the focus should be on stopping their destruction, because natural forests 'are stable, resilient, far better at adapting to changing conditions and store more carbon than young, degraded or plantation forests' (Dooley and Mackey, 2019). A related argument is that the best way to expand the forest area is by allowing degraded forests to naturally recover. 'Allowing trees to regenerate naturally, using nearby remnants of primary forests and seed banks in the soil of recently cleared forests, is more likely to result in a resilient and diverse forest than planting massive numbers of seedlings' (Dooley and Mackey, 2019).

A final consideration is that reforestation in one country should not be achieved through the deforestation of another. The area of forest has been growing in many countries, often through regeneration on abandoned cropland. Table 10.2 records estimates of tree cover change between 1982 and 2016 by climate zone. These estimates are from satellite imagery, and show that there has been growth in net tree cover in all non-tropical zones, and

Table 10.2 Change in world tree cover canopy by climate zone, 1982–2016 (thousand square kilometres)

Climate zone	*Loss*	*Gain*	*Net*
Tropical	927	837	−90
Subtropical	105	448	+343
Temperate	92	951	+859
Boreal	194	723	+529
Polar	7	55	+48
World	1325	3014	+1689

Source: Song et al. (2018), Extended Data Table 2.

especially in the temperate zone where most high-income countries are located. While the table suggests a positive global trend in tree cover, the non-tropical countries that have protected and extended their forest areas have partly done so by importing timber and wood products from tropical countries, where there is continuing to be a net loss of forest cover (see Ritchie and Roser, 2021a).

Proforestation

Proforestation is increasing the area of existing forests that are managed as intact ecosystems in which logging and other extractive industries are prohibited. A study of the potential of proforestation in the USA argued that:

> Intact forests—largely free from human intervention except primarily for trails and hazard removals—are the most carbon-dense and biodiverse terrestrial ecosystems, with additional benefits to society and the economy. Internationally, focus has been on preventing loss of tropical forests, yet U.S. temperate and boreal forests remove sufficient atmospheric CO_2 to reduce national annual net emissions by 11%. U.S. forests have the potential for much more rapid atmospheric CO_2 removal rates and biological carbon sequestration by intact and/or older forests.
>
> (Moomaw et al., 2019, p. 1)

Discussion

The previous two sections of the chapter have explored strategies to increase global tree cover by both reducing future deforestation and reforesting previously cleared land. One research project estimates that the emission of an average of 1.0 gigatonnes of carbon a year could be avoided over the rest of this century by preventing future forest and peatland conversion, while 0.8

gigatonnes a year could be absorbed by reforestation (Turner, 2018, p. 19). This is about 18% of current fossil fuel emissions, a not insignificant amount. A second research project estimates a slightly smaller contribution from both strategies, because it takes a more cautious view of how much land can be reforested without affecting food production and ecosystem integrity. It also argues that it will take at least two decades for forest restoration and reforestation programs to have much effect on carbon sequestration. Both studies agree that these programs are an essential component of policies to restrict future global warming, but have too small an effect to be a substitute for the rapid elimination of fossil fuel emissions. Consequently neither study supports large-scale tree planting. Instead, the second emphasises that: 'We must preserve existing forests and ecosystems, which contain vast quantities of carbon stocks, and continue to remove carbon from the atmosphere while restoring ecosystems and phasing out fossil fuel emissions' (Dooley et al., 2022, p. 819).

These arguments are important because they question the effectiveness of the pledges made by major fossil fuel corporations to become net zero in their carbon emissions. These pledges involve planting large areas of forest to offset continuing fossil fuel use, but have been criticised because the land areas required are unrealistically large and will conflict with other land uses. Furthermore, the ability of planted forests to sequester carbon will take decades to be significant and will decline as they reach maturity.

A second issue is that restoring land cover is not just about managing carbon emissions, because forests and woodland have many other vital functions. However, there are differing views in the research literature on the relative importance of these functions, and therefore on what actions are needed, and where. For example, some environmental groups and international environmental organisations consider that carbon storage is the most important function of forests, because of their concerns about global warming. This results in a focus on tropical forests in the Americas, Southeast Asia and Central Africa, because these have the most carbon-dense vegetation, the largest carbon stocks and the fastest rates of carbon sequestration, and are continuing to be deforested. On the other hand, other researchers argue that:

> The climate-regulating functions of forests—atmospheric moisture production, rainfall and temperature control at local and regional scale—should be recognized as their principal contribution, with carbon storage, timber and non-timber forest products as co-benefits.
>
> ...
>
> This represents a reversal of roles from the current carbon-centric model, where non-carbon effects are treated only as co-benefits.
>
> (Ellison et al., 2017, p. 57)

Sheil provides further support for this viewpoint, with a focus on water.

> Life depends on water while water frequently depends on life. Understanding these dependencies is crucial in ensuring the reliable availability of fresh water. We know that forests and trees play a major role though many details remain debated. In the future, forests should be protected, managed and planted, at least in part, for their role in sustaining atmospheric water and all that depends on it.
>
> (Sheil, 2018, p. 15)

So strategies need to be designed to provide multiple benefits.

> The ubiquity of trade-offs implies that prioritizing a single goal on a land e.g., nature conservation as in the Half-Earth framing, or tree planting as in the 'Trillion Trees Initiative,' would severely impact other functions if these trade-offs are not explicitly taken into account. Using more land for strict, so-called fortress conservation would impact human benefits derived from this land. Maximizing carbon sinks on land through large-scale reforestation or bioenergy production, for instance is unlikely to provide adequate co-benefits for food security, nature conservation, or water provision.
>
> (Meyfroidt et al., 2022, p. 5)

These arguments make land clearing a problem anywhere it is occurring, and not an issue focused only on tropical forests. Land clearing is as much a global issue in the USA and Australia, for example, as it is in Indonesia or Brazil.

A third and very important issue is the effects of some of these strategies on human wellbeing, as they have been criticised as 'carbon colonialism'. This is because controlling deforestation in lower-income tropical countries in order to offset the carbon emissions of high-income countries may limit the opportunities of poor farmers in the former to make a living. So whose wellbeing should take precedence, Amazonian peasants or New York city dwellers; Indonesian timber and plantation workers or middle class Australians? Should the main emphasis be on reducing emissions in high-income countries? Or are tropical peoples also threatened by both climate change and the environmental degradation produced by land clearance? Can forest protection and restoration be combined with the protection of the rights of Indigenous and local communities? Assessment of differing strategies is therefore not just a matter of evaluating the scientific evidence, but also raises issues of human wellbeing and equity. A report for a non-government organisation (Dooley and Stabinsky, 2018) expands this thinking to argue that actions to mitigate climate change must be holistic and also address food security, biodiversity and ecosystem health, and human inequality.

How is geographical thinking used in this chapter?

This chapter has been written to illustrate the application of geographical ways of thinking. As a guide for teachers, the ways in which it does this are summarised in the table below, which lists the main core and subsidiary concepts discussed in earlier chapters of the book that are applied in this one (Table 10.3).

Table 10.3 How the concepts are applied in the chapter

Core concepts *Subsidiary concepts*	*Application in the chapter*
Environment Human alteration of the environment	The whole chapter is about the interrelationships between humans and the environment and illustrates a major component of the meta-concept of environment. The geographer Billie Turner II writes: [The] historical and intellectual antecedents [of land change science] can be found in geography, perhaps more so than any other field of study, and professional geographers make up a substantial proportion of the practitioners in the subfield. (Turner II, 2017) Land change science, more so than any other endeavour, highlights the full expanse of the geographical sciences because of its spatially explicit treatment of the coupled human-environment system. (Turner II, 2009, pp. 173–174)
Interconnection Holistic thinking	The chapter takes a holistic approach to understanding the causes and effects of land cover change, and the choice of strategies to respond to them, using a wide range of research and ideas from across the natural and social sciences. This holistic approach also includes the use of both quantitative and qualitative methods to identify causes. This is the opposite of the reductionist approaches described below by the geographer David Lopez-Carr: Although a large literature now exists on the wide range of demographic, economic, social and ecological processes driving frontier forest conversion, disciplinary rigidity precludes the incorporation into research designs of the full suite of these cross-disciplinary factors. Economists study labour investments and market price fluctuations; demographers investigate fertility, migration and life cycle features; ecologists examine environmental change; and political scientists research institutional processes governing resource use. (Lopez-Carr, 2021, p. 15)
Interconnection Causal relationships	To explain the effects of land clearing on weather and climate the chapter examines the interconnections between the land surface and the atmosphere. To understand the causes of land cover change the chapter argues that it is necessary to establish the causal interconnections, or mechanisms, that produce land clearing.

(*Continued*)

Table 10.3 (Continued)

Core concepts *Subsidiary concepts*	*Application in the chapter*
Place	The causes of land cover change are place-specific, and broadly differ by world region, and at a smaller scale within each world region and country. Lopez-Carr concludes that: Research suggests that efforts to achieve an all-encompassing theory [of frontier deforestation] will continue to pay attention to local context, to the mélange of physical and human geographies unique to each place, and to the fluid nature of space—especially when considering highly mobile human agents of change such as frontier migrants. (Lopez-Carr, 2021, p. 13) Consequently the design of policies to manage land cover change also needs to be place-specific. In a review of REDD+ programs the geographer Andrew McGregor and his colleagues argue that: Current instrumentalist approaches based on abstract models of human behaviour, in this case the use of economic incentives to shift human–forest relations, ignore socioecological histories, diverse place-based motivations and values and pay too little attention to existing social, economic and political contexts. Quite simply, one size does not fit all. For programmes like REDD+ to work, they must emerge from genuine understandings and partnerships involving financiers, practitioners and participants—and they must be flexible enough to respond to different place-based concerns. (McGregor et al., 2015, p. 4)
Space Spatial distribution	The effects of land cover change on climate vary spatially.
Sustainability	The chapter applies the concept of sustainability to evaluate whether too much of the Earth's vegetation has been cleared.
Human wellbeing	The chapter applies the concept of human wellbeing to evaluate the effects of land cover change on people and communities, and the choice of strategies to manage it.

Conclusion

Land cover change is both an old and a very new topic of geographical research. The chapter has argued that land clearing for timber, tree plantations, crops and grazing has resulted in environmental changes that are a threat to environmental sustainability and human wellbeing, as they generally reduce rainfall, increase temperatures, diminish and degrade water supplies, degrade soils, damage biodiversity and reduce the sequestration of carbon. With a high level of confidence, the answer to the question 'Have we cleared too much of the Earth's vegetation' is yes. A variety of strategies are available to reduce land clearing and restore forest cover as a response to this conclusion, but which ones should receive priority, and where, is more debatable, and depends on an assessment of the relative importance of the different ecosystem functions of forests and woodland, the competing uses of land and the wellbeing of different populations. There is much in this for students to discuss and debate.

Summary

- Land cover is the physical and biological cover of the land surface of the Earth.
- Almost half of the habitable surface of the Earth has been converted to cropland and grazing land.
- Land cover changes affect weather and climate by changing surface evapotranspiration, albedo, surface roughness, and net carbon emissions.
- The combined effects of these changes vary with latitude, but mostly result in increased temperatures and reduced rainfall.
- Land cover change can also cause soil erosion and salinity, damage ecosystem services and biodiversity and reduce water quality.
- An application of the concepts of sustainability and human wellbeing suggest that too much of the Earth's vegetation has been cleared.
- The causes of land cover change can be identified though both quantitative and qualitative methods, but only the latter are able to demonstrate the cause-and-effect mechanisms involved.
- Understanding the reasons for land cover change requires a multilevel and holistic approach.
- Strategies to respond to the loss of forest and woodland cover are either designed to prevent further vegetation clearance, or to increase forest and woodland cover by reforestation, afforestation and proforestation.
- The strategies selected need to combine a number of objectives—mitigate climate change, preserve ecosystem services and biodiversity, protect water resources and maintain food and fibre production. They must also be socially just, including ensuring the protection of Indigenous communities.

How could you use this chapter in teaching?

There is a lot in this chapter, and teachers could select portions to suit their program, as long as at some stage they examine strategies to increase forest and woodland. Sections can also be simplified for younger students. Questions to ask your students are:

- Is the question posed at the beginning of the chapter a significant one? If the answer is yes, what makes it significant?
- How can we determine if humans have cleared too much of the Earth's vegetation?
- What strategies would you use to reduce further land clearance, and why?
- What strategies would you use to restore tree cover, and why?
- What projects could you become involved in to restore tree cover? (There are revegetation projects in most urban and rural areas in which students could be personally involved)
- How do we know? (This is a question that can be asked regularly, to encourage thinking about the validity of the information being used and to recognise that knowledge is continually being developed, challenged and improved).
- How are the core concepts being applied in what you are studying?

Useful readings

Antonarakis, A. (2018). Linking carbon and water cycles with forests. *Geography*, *103*(1), 4–11. https://doi.org/10.1080/00167487.2018.12094029

Dooley, K., & Mackey, B. (2019). Want to beat climate change? Protect our natural forests, *The Conversation*, 7 August. https://theconversation.com/want-to-beat-climate-change-protect-our-natural-forests-121491

Ellis, E. C. (2021). Land-use and ecological change: A 12,000-year history. *Annual Review of Environment and Resources*, *46*, 1–33. https://doi.org/10.1146/annurev-environ-012220-010822

Ellis, E. C., Kaplan, J. O., Fuller, D. Q., Vavrus, S., Goldewijk, K. K., & Verburg, P. H. (2013). Used planet: A global history. *PNAS*, *110*(20), 7978–7985. https://doi.org/10.1073/pnas.1217241110

IPCC. (2019). *Climate Change and Land: An IPCC special report on climate change, desertification, land degradation, sustainable land management, food security, and greenhouse gas fluxes in terrestrial ecosystems* (Shukla, P. R. et al. Eds.). https://www.ipcc.ch/srccl/. See chapter 2 on land-climate interactions

Ritchie, H., & Roser, M. (2021a). *Deforestation and forest loss.* OurWorldInData.org. https://ourworldindata.org/forests-and-deforestation

Ritchie, H., & Roser, M. (2021b). *Drivers of deforestation.* OurWorldInData.org. https://ourworldindata.org/drivers-of-deforestation

References

Baker, J. C. A. (2021). Planting trees to combat drought. *Nature Geoscience*, *14*, 458–459. https://doi.org/10.1038/s41561-021-00787-0

Caldas, M., Walker, R., Arima, E., Perz, S., Aldrich, S., & Simmons, C. (2007). Theorizing land cover and land use change: The peasant economy of Amazonian deforestation. *Annals of the Association of American Geographers*, *97*(1), 86–110. https://doi.org/10.1111/j.1467-8306.2007.00525.x

Carrero, G. C., Fearnside, P. M., do Valle, D. R., & de Souza Alves, C. (2020). Deforestation trajectories on a development frontier in the Brazilian Amazon: 35 years of settlement colonization, policy and economic shifts, and land accumulation. *Environmental Management*, *66*, 966–984. https://doi.org/10.1007/s00267-020-01354-w

Curtis, P. G., Slay, C. M., Harris, N. L., Tyukavina, A., & Hansen, M. C. (2018). Classifying drivers of global forest loss. *Science*, *361*, 1108–1111. https://doi.org/10.1126/science.aau3445

DeFries, R. S., Rudel, T., Uriarte, M., & Hansen, M. (2010). Deforestation driven by urban population growth and agricultural trade in the twenty-first century. *Nature Geoscience*, *3*, 178–181. https://doi.org/10.1038/ngeo756

Dooley, K., Nicholls, Z., & Meinshausen, M. (2022). Carbon removals from nature restoration are no substitute for steep emission reductions. *One Earth*, *5*, 812–824. https://doi.org/10.1016/j.oneear.2022.06.002

Dooley, K., & Stabinsky, D. (2018). *Missing pathways to 1.5°C. The role of the land sector in ambitious climate action.* Climate Land Ambition and Rights Alliance. https://www.fern.org/publications-insight/missing-pathways-to-1-5-c-the-role-of-the-land-sector-in-ambitious-climate-action-41/

Ellison, D., Morris, C. E., Locatellie, B., Sheil, D., Cohen, J., Murdiyarso, D., Gutierrez, V., van Noordwijk, M., Creed, I. F., Pokorny, J., Gaveau, D., Spracklen, D. V., Bargués Tobella, A., Ilstedt, U., Teuling, A. J. Gebreyohannis Gebrehiwot, S., Sands, D. C., Muys, B., Verbist, B., ... Sullivan, C. A. (2017). Trees, forests and water: cool insights for a hot world. *Global Environmental Change*, *43*, 51–61. https://doi.org/10.1016/j.gloenvcha.2017.01.002

Fletcher, M. S., Hamilton, R., Dressler, W., & Palmer, L. (2021). Indigenous knowledge and the shackles of wilderness. *PNAS*, *118*(40), e2022218118. https://doi.org/10.1073/pnas.2022218118

Gordon, L., Dunlop, M., & Foran, B. (2003). Land cover change and water vapour flows: Learning from Australia. *Philosophical Transactions of the Royal Society B*, *358*, 1973–1984. https://doi.org/10.1098/rstb.2003.1381

Goudie, A. S. (2019). *Human impact on the natural environment: Past, present and future.* Wiley Blackwell.

Lambin, E. F., & Geist, H. J. (Eds.) (2006). *Land-use and land-cover change: Local processes and global impacts.* Springer.

Lawrence, D., Coe, M., Walker, W., Verchot, L., & Vandecar, K. (2022). The unseen effects of deforestation: Biophysical effects on climate. *Frontiers in Forests and Global Change*, *5*, 756115. https://doi.org/10.3389/ffgc.2022.756115

Leite-Filho, A. T., Soares-Filho, B. S., Davis, J. L., Abrahão, G. M., & Börner, J. (2021). Deforestation reduces rainfall and agricultural revenues in the Brazilian Amazon. *Nature Communications*, *12*, 2591. https://doi.org/10.1038/s41467-021-22840-7

le Polain de Waroux, Y., Garrett, R. D., Chapman, M., Friis, C., Hoelle, J., Hodel, L., Hopping, K., & Zaehringer, J. G. (2021). The role of culture in land system science. *Journal of Land Use Science*, *16*(4), 450–466. https://doi.org/10.1080/1747423X.2021.1950229

Li, X., Tian, H, Pan, S., & Lu, C. (2022). Four-century history of land transformation by humans in the United States: 1630–2020. *Earth System Science Data Discussions* [preprint]. https://doi.org/10.5194/essd-15-1005-2023

Lopez-Carr, D. A. (2021). A review of small farmer land use and deforestation in tropical forest frontiers. *Land*, *10*, 1113. https://doi.org/10.3390/land10111113

McCarthy, S., & Tacconi, L. (2011). The political economy of tropical deforestation: Assessing models and motives. *Environmental Politics*, *20*, 115–132. https://doi.org/10.1080/09644016.2011.538171

McGregor, A., Eilenberg, M., & Coutinho, J. B. (2015). From global policy to local politics: The social dynamics of REDD+ in Asia Pacific. *Asia Pacific Viewpoint*, *56*(1), 1–5. https://doi.org/10.1080/09644016.2011.538171

Meyfroidt, P., de Bremond, A., Ryan, C. M., Archer, E., Aspinall, R., Chhabra, A., Camara, G., Corbera, E., DeFries, R., Diaz, S., Dongo, J., Ellis, E. C., Erb, K-H., Fisher, J. A., Garrett, R. D., Golubiewski, N. E., Grau, H. R., Grove, M., Haberl, H., … zu Ermgassen, E. K. H. J. (2022). Ten facts about land systems for sustainability. *PNAS*, *119*(7), e 2109217118, 1–12. https://doi.org/10.1073/pnas.2109217118

Moomaw, W. R., Masino, S. A., & Falson, E. K. (2019). Intact forests in the United States: Proforestation mitigates climate change and serves the greatest good. *Frontiers in Forests and Global Change*, *2*, Article 27. https://doi.org/10.3389/ffgc.2019.00027

Seymour, F., & Harris, N. L (2019). Reducing tropical deforestation. *Science*, *365*(6455), 756–757. https://doi.org/10.1126/science.aax8546

Sheil, D. (2018). Forests, atmospheric water and an uncertain future: The new biology of the global water cycle. *Forest Ecosystems*, *5*(19). https://doi.org/10.1186/s40663-018-0138-y

Song, X-P., Hansen, M. C., Stehman, S. V., Potapov, P. V., Tyukavina, A., Vermote, E. F., & Townshend, J. R. (2018). Global land change from 1982 to 2016. *Nature*, *560* (7720), 639–643. https://doi.org/10.1038/s41586-018-0411-9

Turner II, B. L. (2009). Land change (systems) science. In N. Castree, D. Demeritt, D. Liverman, & B. Rhoads (Eds.), *A companion to environmental geography* (pp. 168–180). Wiley-Blackwell.

Turner II, B. L. (2017). Land change science. In D. Richardson, N. Castree, M. F. Goodchild, A. Kobayashi, W. Liu, & R. A. Marston (Eds.), *The International Encyclopedia of Geography*. John Wiley. https://onlinelibrary.wiley.com/doi/10.1002/9781118786352.wbieg0192

Turner II, B. L., Meyfroidt, P., Kuemmerlee, T., Müllere, D., & Chowdhury, R. (2020). Framing the search for a theory of land use. *Journal of Land Use Science*, *15*(4), 489–508. https://doi.org/10.1080/1747423X.2020.1811792

Turner, W. R. (2018). Looking to nature for solutions. *Nature Climate Change*, *8*, 14–21. https://doi.org/10.1038/s41558-017-0048-y

Veldman, J. W., Overbeck, G. E., Negreiros, D., Mahy, G., Le Stradic, S., Fernandes, G. W, Durigan, G., Buisson, E., Putz, F. E., & Bond, W. J. (2015). Where tree planting and forest expansion are bad for biodiversity and ecosystem services. *BioScience*, *65*(10), 1011–1018. https://doi.org/10.1093/biosci/biv118

Wilderness Society. (2022). *Watch on nature*. Wilderness Society. https://www.wilderness.org.au/protecting-nature/watch-on-nature

Williams, M. (2006). *Deforesting the earth: From prehistory to global crisis: An abridgement*. University of Chicago Press.

Chapter 11

Unequal places

Should governments try to reduce regional inequalities?

The topic is expressed as a question to be debated; not as content to be learned.

> In 2014 world leaders ranging from presidents of America and China, the Pope, the Business leaders who meet at Davos and even those who now run the World bank all ranked economic inequality as an issue of importance among the top three in the world.
>
> (Dorling, 2015, p. 24)

The quotation identifies economic inequality as a major contemporary issue. It can be studied at an international, national or urban scale; the first of these is a common topic in school geography, often under the title of development geography. At the national scale, inequality can be examined in two ways: one looks at the differences in incomes within a country between richer and poorer groups of people, which is more common, while the other looks at income differences between regions within a country. Both forms of inequality contribute to the issues identified in the quotation, but this chapter takes the second more geographical approach to understanding economic inequality and its consequences by focusing on regional inequality. This is a topic sometimes missing from school curriculums, but it is one that affects the lives of many students in Europe, the USA, Canada, Australia and New Zealand, and is associated with some significant, economic, social, environmental and political issues. To keep it manageable, the discussion is limited to high-income countries.

To answer the question, students will first need to understand what economic inequality means and looks like and then examine regional inequality. They then will need to identify the effects of regional economic inequality and decide if these effects justify a policy response. On the principle that young people should never learn about a problem without also learning about how it could be mitigated, the chapter also reviews ways of reducing regional inequalities.

DOI: 10.4324/9781003376668-12

First step: The first step in investigating this question is to understand what economic inequality is, how it is measured, and how it has been changing in different countries.

What is economic inequality?

Economic inequality is the difference between people in their income (their earnings), and their wealth (their accumulated assets), but for simplicity, this chapter mostly only examines income. Inequality in income can be measured in several ways, such as by:

- A comparison of the share of total national income gained by the top 10% of individuals with that of the bottom 50%.
- The ratio of the average income of the top 10% of income earners to the average income of the bottom 50%. This measure represents how many times more the average income of the top 10% is compared with the bottom 50%. For example, in Table 11.1, the average income of the top 10% of adults in Sweden is seven times those in the bottom 50%, whereas in the USA it is 17 times.
- The share of total national income of the top 1% of income earners. Many writers on inequality regard this as the most revealing statistic, as it measures the extent to which income distribution is skewed towards the extremely rich.

To illustrate the extent of economic inequality, Table 11.1 shows these indicators for a range of countries, in order of increasing inequality as measured by the 10/50 income ratio. Sweden is an example of a country with a relatively low level of income inequality, and the USA an example of relatively high

Table 11.1 Income and wealth inequality for selected countries, 2021

Country	*10/50 income ratio*	*Top 10%*		*Bottom 50%*		*Top 1%*	
		Income	*Wealth*	*Income*	*Wealth*	*Income*	*Wealth*
Sweden	7/1	32.0	58.9	23.3	4.8	11.7	27.7
UK	9/1	35.8	57.1	20.3	4.6	12.7	21.3
Germany	10/1	37.8	58.9	18.6	3.4	13.3	28.6
Australia	10/1	32.6	57.1	16.6	5.9	11.3	23.9
Canada	12/1	39.7	58.3	16.3	4.8	13.9	24.9
Japan	13/1	44.2	58.6	16.8	4.8	12.9	24.8
USA	17/1	45.6	70.7	13.8	1.5	19.1	34.9
South Africa	56/1	65.4	85.6	5.8	−2.5	19.3	54.9

Source: World Inequality Database (https://wid.world/data/).

inequality for a high-income country. South Africa is an example of extreme inequality, where the difference between the top 10% and bottom 50% has increased since the end of apartheid in 1991, and the wealth of the bottom 50% is negative because the debts of this group are estimated to be bigger than their assets. The table also include estimates of the distribution of wealth, which show that it is much more unequally distributed than income.

What the table also reveals is that countries with similar average income levels can have different degrees of inequality because, as the World Inequality Report 2022, explains

> The degree of inequality within a society is fundamentally a result of political choices: it is determined by how a society decides to organize its economy (i.e. the sets of rights given to and constraints imposed on firms, governments, individuals, and other economic actors).
>
> (Chancel et al., 2022, p. 30)

Trends in income inequality

Inequality in all countries has changed over time. This is illustrated for Sweden, Australia and the USA in Figures 11.1, 11.2 and 11.3. The graphs not only show the extent of inequality, as measured by the gap between the top 10% and bottom 50%, but also how this has changed since 1980. The trends in inequality in each country are described below in extracts from the World Inequality Report 2022.

> Sweden was one of the most unequal countries in Europe in the late 19th and early 20th centuries and democratic rights were tied to wealth ownership. The expansion of democracy and growing support for the Swedish socialist party paved the way for the development of the Swedish welfare state, which led to a large-scale drop in inequalities, accompanied by fast-rising average incomes for the vast majority of the population. While inequalities have risen in Sweden since the 1980s … the country remains one of the most equal nations on earth in 2021.
>
> (Chancel et al., 2022, Country Sheets)

> Income inequality in Australia has been rising steadily since the early 1980s. … This trend stands in contrast with the 1900–1970 period, when top incomes experienced a severe drop, making 1970s Australia one of the most equal countries on the planet. The 2010s has been a lost decade, in particular for the bottom 50%, whose incomes are slightly under their level 10 years ago.
>
> (Chancel et al., 2022, Country Sheets)

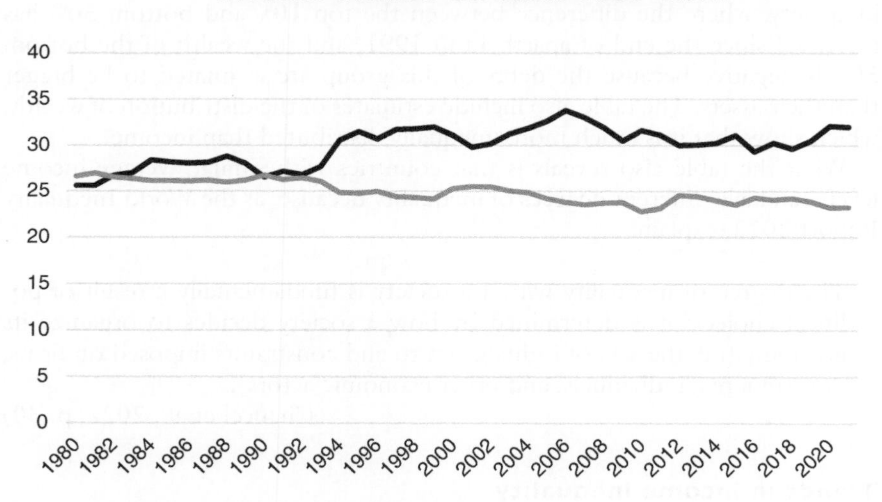

Figure 11.1 Income shares in Sweden, 1980–2021.

The top line is the income share of the top 10%, and the bottom line the income share of the bottom 50%.

Source: World Inequality Database https://wid.world/data/.

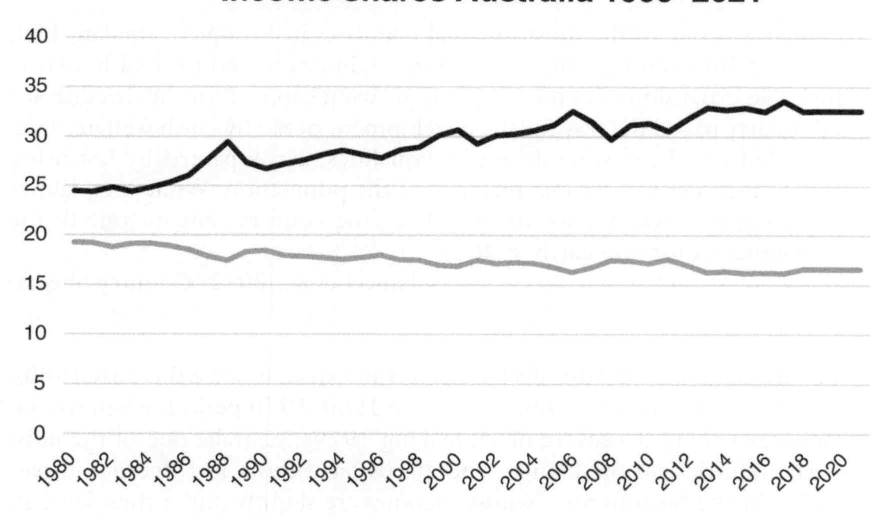

Figure 11.2 Income shares in Australia, 1980–2021.

The top line is the income share of the top 10%, and the bottom line the income share of the bottom 50%.

Source: World Inequality Database https://wid.world/data/.

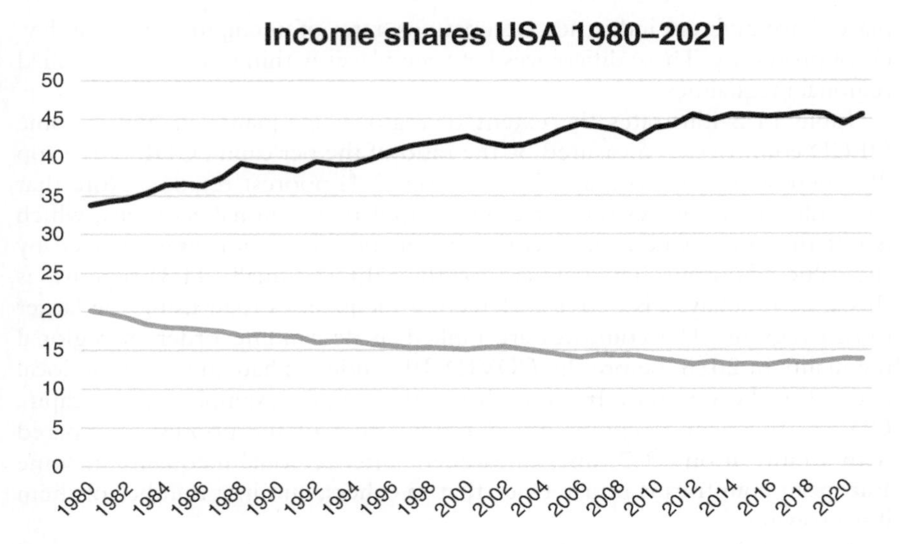

Figure 11.3 Income shares in the USA, 1980–2021.

The top line is the income share of the top 10%, and the bottom line the income share of the bottom 50%.

Source: World Inequality Database https://wid.world/data/.

> The top 10% income share in the US dropped significantly after the Second World War … under the effect of strong capital control policies and a rise in federal spending, accompanied by strongly progressive taxation. The 1950–1980s were also marked by rapidly rising average incomes. From the early 1980s onward, deregulation, privatizations, decreases in tax progressivity and a decline in union coverage all contributed to a formidable rise in the top 10% income share ….
>
> (Chancel et al., 2022, Country Sheets)

The graphs show that the inequality gap has remained much the same in Sweden for the last two decades, but has widened in both Australia and the USA over the last five decades. This illustrates the point made earlier that the degree of inequality within a country is fundamentally a consequence of political choices.

Next step: What is regional economic inequality?

Regional inequality

People's incomes depend mainly on their occupations, skills, the industries they work in, their gender and race, whether they are employed and other personal characteristics. They are also influenced by where people live, as different places have different industries, occupations, rates of unemployment, and levels of prosperity. These differences between places within a country are called regional inequalities.

Table 11.2 illustrates the extent of regional inequality in high-income OECD countries, as measured by the ratio of the per capita GDP of the top 20% richest regions to that of the bottom 20% poorest regions. Note that this indicator measures the per capita output of a regional economy, which is not the same as personal income, but income is obviously influenced by the value of output. The data are for the OECD's small (TL3) regions, as this scale of analysis is better at detecting inequalities than using the larger (TL2) regions. The countries are ranked in descending order of regional inequality in 2018, before the COVID-19 pandemic had an impact on local and national economies. It shows that in the UK, for example, the per capita GDP of the richest regions was 3.1 times that of the poorest, compared with a ratio of only 1.7 times in Sweden. Inter-regional inequality in some European countries is nearly twice that in others, and in a number of them it is growing.

Table 11.2 identifies the UK as one of the most inter-regionally unequal countries in the OECD, and Table 11.3 illustrates this with data for each

Table 11.2 Regional disparity in selected OECD countries, 2008 and 2018

Top 20%/bottom 20% ratio[a]	*2008*	*2018*	*Change between 2008 and 2018*
Hungary	3.6	3.2	Decrease
UK	2.9	3.1	Increase
Poland	2.7	3.0	Increase
Germany	2.8	2.7	Decrease
France	2.3	2.5	Increase
Belgium	2.6	2.5	Decrease
Greece	2.5	2.5	No change
Italy	2.4	2.5	Increase
Netherlands	2.0	2.1	Increase
Denmark	2.0	2.1	Increase
Japan	2.1	2.0	Decrease
Norway	2.0	2.0	No change
USA	1.8	2.0	Increase
Spain	1.8	1.8	No change
Portugal	2.1	1.8	Decrease
Austria	2.0	1.8	Decrease
Sweden	1.7	1.7	No change

Source: OECD *Regions and cities at a glance*. https://doi.org/10.1787/888934189735.

a. The ratio of the top 20% richest regions over the bottom 20% poorest regions, for OECD small (TL3) regions.

Table 11.3 Regional economic statistics for the UK, 2019

Region	*Regional GDP per capita as a % of Greater London, 2019*	*Growth in GDP at constant prices as % of Greater London, 2005–2019*	*Per capita disposable household income as % of Greater London, 2019*
North East England	42.2	3.5	56.9
North West England	51.2	23.1	60.0
Yorkshire and The Humber	46.7	12.4	59.7
East Midlands	47.0	15.5	62.0
West Midlands	48.0	18.4	61.1
East of England	53.0	22.3	74.2
Greater London	100	100	100
South East England	62.3	42.4	81.9
South West England	51.1	19.3	70.4
Wales	43.5	8.6	57.5
Scotland	53.1	17.1	65.2
Northern Ireland	46	4.7	57.7

Source: OECD. https://stats.oecd.org/viewhtml.aspx?datasetcode=REGION_ECONOM&lang=en#.

region within the country. These reveal the contrast between Greater London and the South East of England, on the one hand, and the rest of the country, on the other. The first column of data compares the GDP per capita of each region in 2019 with that of Greater London and shows that most regions produced only around half the per capita economic output of the London region. The second data column compares the rate of growth in GDP in each region between 2005 and 2019, at constant prices, compared with that of Greater London, and shows that London and the South East have been growing much faster than the rest of the country. Combined, these data indicate that regional inequalities in the UK are large and growing. The last column in the table compares the per capita disposable household income in each region with that of London. This shows that the disparities in disposable incomes between the regions are less that the disparities in their per capita GDP, because of higher taxation in the richer regions and higher government welfare payments in the poorer regions. This role of taxation and welfare payments in redistributing income between regions is an important process, but not always recognised.

The regions in Table 11.3 are quite large and conceal a lot of the difference that exists at finer scales. Figure 11.4 maps weekly earnings in April 2017 by place of work for 380 local authority districts in England, Scotland and Wales and shows a far more complex pattern. Even so, the contrast between the southeast of England and the rest of the country is still apparent, with much of Wales and the northeast of England having relatively low weekly earnings, and London having the highest.

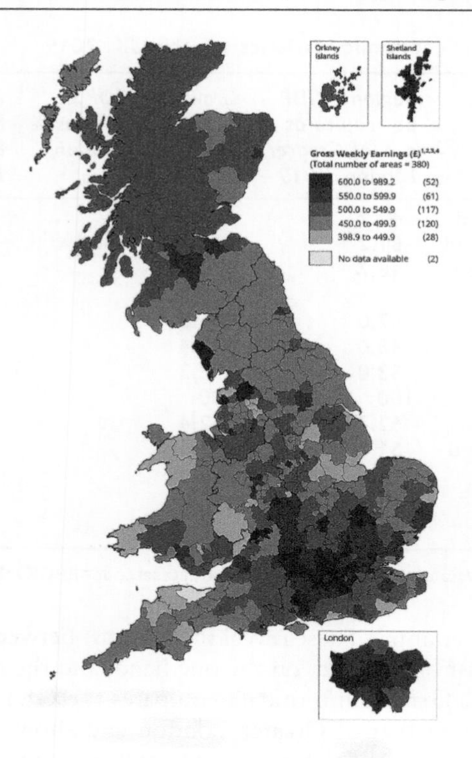

Figure 11.4 Median full-time gross weekly earnings by place of work, Great Britain, 2017.

Source: UK Office for National Statistics https://www.ons.gov.uk/employmentandlabourmarket/peopleinwork/earningsandworkinghours/bulletins/annualsurveyofhoursandearnings/2017provisionaland2016revisedresults.

Should the regional inequalities revealed in these tables and maps be reduced? To do this, policies would have to focus on the poorer places, to raise their incomes, as there is no point in reducing the incomes of the richer places. So the discussion now turns to identifying the most unequal places within high-income countries.

Next step: Which places are the most unequal?

Left behind places

In many advanced economies regional inequalities have tended to widen over the last four decades, and:

> major gaps have opened up between successful places that have 'pulled ahead' economically and places that have failed to share in the same success ... In the period since the 2008 [financial] crash, these latter areas have become characterized in both academic studies and political discourse as 'left behind places'. This is an international phenomenon, with national and subnational particularities, for example, 'La France périphéri-que' (peripheral France), 'abgehängte Regionen' (suspended regions) in Germany, 'Aree Interne' (inner areas) in Italy, 'Krinnpgebieden' (shrinking areas) in the Netherlands, 'la España vaciada' (the hollowed out Spain), and 'legacy cities' and the 'rustbelt' in the United States.
>
> (Martin et al., 2021, p. 12)

In the UK, the economy of London has grown much more than that of other cities. In the USA, a group of 'super star' cities has attracted new technologies and skilled workers, while many former manufacturing cities and towns have lost jobs and population. Hendriksen, Muro and Galston (2018, p. 4) describe the emergence of 'two Americas', 'one based in large, diverse, thriving metropolitan regions; the other found in more homogeneous small towns and rural areas struggling under the weight of economic stagnation and social decline'. However, it is essential to recognise that within economically successful cities, there are major inequalities, such as boroughs in London with low-income households who have not shared in the prosperity of the region, and low-income areas within large American cities, often the result of racial discrimination. These urban areas have similar characteristics to other economically poorer places.

The characteristics that identify left behind places include:

> relative economic under-performance and decline, expressed in below average pay, employment and productivity; lower levels of educational qualifications and skills; higher levels of poverty and disadvantage (compared to national averages); out-migration, ageing and demographic shrinkage; poor health; limited connectivity and investment in social and economic infrastructure; reduced service provision; political disengagement, neglect and discontent; and a lack of civic assets and community facilities. While not all of these specific characteristics will be evident in every 'left behind' place, it is the combination of economic disadvantage, lower living standards, population loss/contraction/ low-growth, a lack of infrastructure and political neglect and disengagement that can be said to define a place as 'left behind'.
>
> (MacKinnon et al., 2022, pp. 41–42)

If there is a case for reducing regional inequalities, these are the places on which policies should focus. So is there a case?

Next step: Should regional inequalities be reduced?

Is there a case for improving the economies of left behind places?

To decide whether governments should be concerned enough about the economies of left behind places to adopt stronger policies to assist them, students should evaluate the arguments about whether this would be beneficial. The arguments, both for and against, are outlined below.

Many economists and government economic agencies, particularly in the UK and US, argue that policies implemented to help lagging regions will reduce national economic growth. This is because they believe that large agglomerations are more efficient, assist the development of new industries and are the main engine of national economic growth. Funds spent on left behind places will not have the same impact on the economy, so there is a 'trade-off' between greater social equity and lower economic growth.

This belief has been widely criticised, for at least four reasons. One is that the apparent greater efficiency of large agglomerations, as discussed in Chapter 5, does not take into account the costs produced by their growth, such as expensive infrastructure, higher housing costs, long journeys to work, social issues and environmental pollution. The second is that not all cities have been successful, as the contrast between Montreal and Toronto, and Birmingham and London, demonstrates. The third is that strategies that focus on the big cities leave large regions of a country with underutilised economic potential. Ron Martin and his colleagues argue that 'there are strong instrumental arguments for reducing spatial inequalities, namely that it increases the utilization and productive potential of the human and business resources in "left behind places" and maximizes their contribution to national growth and development' (Martin et al., 2021, p. 23). A similar point has been made for the USA:

> Leaving struggling places to fend for themselves may reduce public outlays today, only to increase them tomorrow as the consequences of neglect manifest themselves in increased costs for health care, disability, and substance abuse programs. Lower labor force participation, moreover, will restrict prospects for economic growth, a trend that will prove increasingly damaging as our population ages.
>
> (Hendricksen et al., 2018, p. 6)

The fourth reason is that there is little or no empirical evidence that greater regional equality reduces national economic growth. The argument that it does comes from modelling, not from real-world data. Martin and his colleagues conclude that:

> There is in fact little persuasive evidence of a tradeoff between equality and efficiency; on the contrary, the evidence suggests that those countries with lower levels of social and spatial inequality tend to grow just as fast, if not faster, over the long run than those with greater disparities.
>
> (Martin et al., 2021, p. 114)[1]

Part of the efficiency argument is the contention that the growth of the dynamic cities will help the poorer places through the spread of knowledge and skills to them and through the migration of people from them to the cities. Critics of this viewpoint contend that the economic growth of the dynamic cities has not in fact 'trickled down' to the rest of the country. In the UK, for example, the London economy is largely unconnected to the rest of the country and is much more tied into the global economy. Note that this point illustrates the value of the concept of interconnection, which draws attention to the significance of the relationships between places and regions. The belief that migration from left behind places to those with greater opportunity will raise incomes in the former has also been contested. One criticism is that to the extent that migration does happen, it weakens the human resources of the left behind places and adds to those of the advantaged places, with the effect of increasing the economic divergence between them. A second is that this migration is not happening, as inter-regional migration has declined in both the UK and the USA, and has remained relatively low in Europe over the last three decades. In the UK, migration from the North of the country to London and the South has 'essentially come to a halt' (Beatty and Fothergill, 2020, p. 1247). This lack of migration is because many people are unable or unwilling to move. Those who can move have probably done so already, and those who remain lack the skills and qualifications required in the new economic sectors or cannot afford the cost of accommodation in the cities. Many also have emotional attachments to their home place and, in the USA, to local cultural values that are different to those in the big cities. (Ulrich-Schad and Duncan, 2018) Mobility is therefore not costless.

Fiona Hill, who grew up in what she describes as the UK's equivalent of America's Rust Belt, the North East of England, and left it for a distinguished career in the USA, explains why many did not leave the region. In her book, *There is nothing for you here* (which is what her father said to her), she writes:

> The people who stayed in Bishop Auckland [her home town] did so for the same reasons that anyone gives for wanting to remain at home. They wanted to be where their grandparents and parents had put down roots. They didn't want to be transplants in some other place. They wanted a job and the feeling that they were living somewhere that mattered to them—in

> a tight-knit community where they belonged, where people knew their names, and where no one was going to make fun of their accents.
>
> (Hill, 2021, p. 330)

If growth does not spread to left behind places, and migration from them is limited, then the efficiency argument for focusing on the cities leaves them dependent on income transfers through welfare payments. Students could discuss whether this is equitable. People in left behind places appear to have decided that it isn't, as their political reaction has been described as a 'geography of discontent', or 'the revenge of places that don't matter'. Andrés Rodríguez-Pose (2018, p. 200) writes that these areas, which have 'witnessed long periods of decline, migration and brain drain, … have seen better times and remember them with nostalgia, … [and] have been repeatedly told that the future lays elsewhere, have used the ballot box as their weapon'. In the UK they tended to vote to leave the EU, in Europe they tended to vote for parties opposed to the EU, and in the USA they voted for Donald Trump as President. This has made left behind places an issue of considerable political and academic interest. It has also presented puzzles, such as in the UK where the places that voted to leave the EU tend to be ones most dependent on trade with Europe, and on regional development assistance from the EU. London, on the other hand, which voted to remain, is less dependent on Europe as it has a much more global economy. (Los et al., 2017)

There are several other arguments that students could consider. These are based on studies of the effects of higher levels of inequality between people within a country and not on inequalities between regions within a country. However, if the incomes of people in economically poorer places are raised, this will reduce national inequality and reduce the negative effects that research on national inequalities has found. These are outlined below.

One argument is that the degree of inequality within a country has an influence on human wellbeing. This influence has been analysed by Richard Wilkinson and Kate Pickett, two epidemiologists, in their books *The spirit level* and *The inner level* (Wilkinson and Pickett, 2009 and 2018). Andrew Sayer summarises their findings as follows:

> [Wilkinson and Pickett] find that once average incomes in countries reach a certain point, further increases make little difference to well-being as measured by indicators of these things. What does make a difference to well-being in rich countries is the degree of inequality: the more equal of the rich societies tend to score better than the others on a wide range of indicators of well-being, from life expectancy and health, to low crime rates, high educational performance, social mobility and social trust …
>
> (Sayer, 2016, p. 312)

Inequality within a country has also been found to be associated with racism, homophobia, sexism, antagonism towards minorities, opposition to social welfare, belief in conspiracy theories and the volume of online hate tweets (Denti and Faggian, 2021). The cause of this relationship between greater inequality, and poorer wellbeing and negative social attitudes, is thought to be the effects of inequality on the quality of social networks. 'When a person's relationships to coworkers, neighbours, friends, and others weakens—and when lower income brings lower status, isolation, and economic anxiety—a number of psychopathologies and social problems can also take hold' (Soknes, 2021, p. 151).

A second argument is that inequality increases the pressures on the environment. For example, Wilkinson and Pickett write that:

> There are powerful links between inequality, the threat to the environment, and the failure to achieve genuinely higher levels of well-being. Most obviously, greater inequality intensifies consumerism and status consumption. Starker material differences magnify status differences and make us more prone to worries about the impression we create in the minds of others ... Instead of being a source of well-being and fulfilment, as advertising would have us believe, psychological studies confirm that consumerism is driven by status insecurity.
>
> (Wilkinson and Pickett, 2018, p. 228)

Hamann et al. come to a similar conclusion:

> Inequality ... can be a potent driver of behavior in combination with an individual's aspirations to adhere to social norms, achieve goals, or emulate peers who are viewed as more successful. ... For example, if the possession of material goods is highly valued ..., the less successful will try to follow the consumption patterns of the successful. This behavior reinforces the value placed on material goods, driving the successful toward ever higher consumption patterns in order to distinguish themselves from the less successful—thus exacerbating the cycle of environmentally damaging overconsumption.
>
> (Hamann et al., 2018, p. 70)

Similarly, Wilkinson and Pickett believe that:

> If the world is to move towards an environmentally sustainable way of life, it means acting on the basis of the common good as never before, indeed acting for the good of humanity as a whole. But the status insecurity and individualism promoted by inequality distance us from both the means and the will to take action on problems that threaten us all.
>
> (Wilkinson and Pickett, 2018, p. 228)

Inequality is also an economic problem. 'While the relationship between inequality and economic growth is not clear-cut, recent research shows that countries with high and rising inequalities generally experience slower growth than those with lower inequality' (United Nations Department of Economic and Social Affairs, p. 45). The causes of this relationship include:

- Concentrating the power of the wealthy may weaken the support for needed reforms. For example, 'big corporations and the wealthy may use their position and resources to lobby in support of their interests, raise legal challenges to progressive tax legislation, or promote … campaigns to influence … public perceptions of redistribution' (United Nations Department of Economic and Social Affairs, 2020, p. 45).
- Low-income regions within a country, with below average productivity and underutilisation of labour and capital, reduce national economic performance.
- Barriers to the educational attainment of poorer people limit their ability to contribute to the economy.

As an illustration of the economic effects of increasing inequality, an IMF Staff Discussion Note concludes that:

> if the income share of the top 20 percent (the rich) increases, then GDP growth actually declines over the medium term, suggesting that the benefits do not trickle down. In contrast, an increase in the income share of the bottom 20 percent (the poor) is associated with higher GDP growth.
> (Dabla-Norris et al., 2015, p. 4)

The reason for this is that low-income people spend a higher proportion of any increase in income than high-income people.

Perhaps the strongest argument is that inequality has political consequences if it reduces people's trust in the capacity of the political system to address their needs, as 'higher levels of economic inequality have been linked with lower political participation, reduced support for democracy, greater endorsement of authoritarian values and increased support for strong leaders who are willing to use undemocratic means to achieve desired outcomes' (Jetten et al., 2022, p. 1). One form of this political response was noted earlier as a reaction of the left behind places.

These arguments make economic inequality not just an ethical issue of social justice and unfairness but also a contributor to some significant social, environmental and economic problems. From this, students might decide that, while inequality can never be eradicated, it should be reduced, and that if reductions in regional economic inequalities can contribute to this, then there are additional reasons for policies to achieve this outcome. What policies could be tried?

Next step: Students must understand the causes of places being left behind before they can evaluate responses to the problem.

Why are places being left behind?

As argued before in this book, understanding the causes of a problem is essential before responses can be identified. This section briefly describes the major causes of the economic problems of left behind places.

The expansion of global trade towards the end of the last century, and particularly the rise of China as an exporter of manufactured products, led to the closure of manufacturing plants, particularly in towns in the UK and USA. How much employment was lost in the USA is unclear, as different economists have produced different estimates, although they tend to agree that job losses caused by trade with China ceased after 2010. Tellingly, however

> voters in [US] congressional districts exposed to greater increases in Chinese imports disproportionately removed moderate politicians from office in the 2000s, … [i]n the United Kingdom, regions exposed to greater inflows of Chinese goods voted to leave the European Union at higher rates in the 2016 referendum, [a]nd across 15 European countries, geographically concentrated import shocks are associated with higher vote shares for nationalist, isolationist, and radical right parties.
>
> (Rickard, 2020, p. 194)

The economic liberalisation policies followed after 1980 by many Western countries added to regional economic divergence. For example, in the USA deregulation of airlines and banking, and reduced control of company mergers and takeovers, disadvantaged smaller cities and towns because they ended up with poorer airline services, loss of the community banks that supported local businesses and loss of local entrepreneurs (Hendriksen, Muro and Galston, 2018). Similarly, in the UK, the deregulation of the financial and banking system in 1986 provided opportunities for London to become a major global financial centre.

Technological innovations created new industries, often in new places, reduced employment in old industries in old places and made outsourcing of production to other countries cheaper. In the USA, new industries based on digital technologies developed in or were attracted to cities that were mostly not old industrial centres, such as Silicon Valley. Sandbu writes that 'the economic transformation [of the USA] of the past forty years has shifted the engine of value creation from a territorially spread out system of labour-intensive industrial production to more concentrated activity of knowledge-based and high-tech services' (quoted in Martin et al., 2021, p. 55). Similarly, for the UK, Martin and his colleagues conclude that

> The primary cause of this process of spatial divergence lies in the inability of some places to adapt to the growth of the post-industrial service and knowledge-based economy whose locational requirements are very different from those of past heavy industries.
>
> (Martin et al., 2021, p. 107)

For Europe, Andrés Rodríguez-Pose, Michael Storper and Simona Iammarino explain that:

> The major wave of technological innovation that began in the 1970s has stimulated the concentration of high-technology and knowledge-intensive sectors in large metropolitan areas, favouring the mobility of highly skilled, non-routine and creative jobs towards economic cores. The increasing automation of previously dominant manufacturing industries has revolutionised trade costs and resulted in the substitution of routinised medium- and low-skilled jobs in most of the former industrial hubs of Europe. Manufacturing activity has become more geographically dispersed – and increasingly outsourced to third countries – leading to the demise of the more routine industry jobs across most of Europe.
>
> (Rodríguez-Pose et al., 2018, n.p.)

Economic policies in the UK and USA have favoured the large cities. Danny MacKinnon and his colleagues explain why, in ways that will be familiar to those who have read Chapter 5:

> over the past couple of decades, cities have been identified as key engines of economic growth and innovation by academics, governments and international economic organisations. Based on the NEG [New Economic Geography] and urban economics, the underlying argument is that the geographical agglomeration of economic activity in cities fosters innovation and productivity gains, as concentrations of firms and skilled workers generate knowledge spill-overs. According to Rodríguez-Pose, this 'dominant narrative' 'puts forward the idea that big cities are the future and that the best form of territorial innovation is not to focus on declining places – perceived as having low potential – but to bet on what is perceived to be the winning horse: the largest and most dynamic agglomerations.
>
> (MacKinnon et al., 2022, p. 43)

An example of this preference for large cities is the way that UK government investment in transport infrastructure has favoured London, with average annual per capita investment in transport infrastructure 2.8 times higher in London that in the remainder of the country between 2014–2015 and 2018–2019 (Harris and Moffat, 2022, p. 1725). In addition, UK government policies that favoured the financial and business services sectors over manufacturing further advantaged London and the South East of England and neglected much of the rest of the country. The growth of London's finance sector has also had negative effects on the rest of the country, by diverting resources from non-financial activities such as manufacturing, concentrating investment in London and the South East, drawing human and financial capital from the regions and dominating government policy-making. Doreen Massey drew attention to the unequal power relationships between London and the rest of the country over two decades ago (Massey, 2001, p. 7), while a recent article argues that:

> the problem of regional economic inequality in the UK, and especially the falling behind of much of northern Britain, cannot be understood in isolation from the concentration of economic, financial and political power in London, and the strategic policy priorities of the state, itself overwhelmingly centralised in London.
>
> (Martin and Sunley, in press)

Investment in transport connections between lower-income regions and major centres has not always helped the former, because, for example, 'a high-speed train line between two very unequal territories often reinforces centralisation and can lead to de-industrialisation, fewer locally provided services, and a decline in local commerce' (Iammarino et al., 2019, p. 286).

People's attitudes may differ according to the economic condition of the region in which they live, in ways that perpetuate these differences. For example, McCann (2020, p. 257) reports that 'Numerous social surveys demonstrate that people whose life is primarily in prosperous regions tend to have a profoundly different view of the world, themselves and their opportunities for self-enhancement than those who live in low-productivity regions'.

A further obstacle to the redevelopment of left behind places is that they are often dependent on toxic industries, such as energy generation from coal, the processing of metals and the production of chemicals, and have degraded and unhealthy environments. Bez and Virgillito, in their study of toxic pollution and labour markets in Europe, argue that these places are locked into a dependence on old industries and forced to accept a trade-off between health and employment. This discourages investment in new and cleaner industries, which locate elsewhere. They found that 'regions where fewer toxic pollutants are emitted are regions with in-migration flows, while the opposite is true for regions characterised by a highly diversified, highly polluting mix of pollutants' (Bez and Virgillito, 2022, pp. 40–41). The environment is therefore another contributor to the problems of the left behind places.

However, it must be recognised that there are examples of left behind cities and towns that have successfully developed new, high value-added activities, often by evolving out of former older industries, or by new strategic initiatives based on developing and building on local science-based universities, purposive institutions and collaborative endeavours. Some of these are places in what became known as the US Rust Belt (R. Martin, personal communication).

If these are the causes of growing regional inequalities in high-income countries, what are the implications for policies to address these inequalities?

Last step: Students can now evaluate the strategies that can be tried to improve the economies of left behind places.

Policies for left behind places

The discussion in the previous section of the causes of regional divergence and the consequences for left behind places strongly suggests that past economic policies, some of which were only concerned with the national economy, and others which focused on efficiency through agglomeration, are unlikely to reduce regional inequalities. In fact, they are more likely to increase them. Instead, Martin and his colleagues argue that:

> A major corrective to [economic policymaking that does not] take the country's regional and urban structure into account … is required. It is in individual cities, towns and localities that the 'everyday business of economic life' is conducted: where production is carried out, exports originate, wages are earned and partly spent, and social and public services are provided and accessed. The spatial structure and organization of the economy is not some incidental feature but foundational to how well the national economy performs, functions and prospers: geography matters. A new 'spatial imaginary' needs to become an integral feature of political economic thinking and national economic management.
>
> (Martin et al., 2021, p. 113)

In a number of countries, academics and policy analysts are now advocating alternative strategies that are geographical. These can take two forms. One is described as place-sensitive. This groups similar places into types based on their per capita incomes and develops strategies that address the strengths and weaknesses of each type. In this they are sensitive to regional differences. However, they tend to be centrally directed strategies that may be insufficiently aligned with the needs of individual places or provide insufficient scope for local discretion or control. The other approach is called place-based. This involves:

> Policies and strategies designed and implemented by a place's local authority and other local policy actors and institutions, intended to respond to the specific problems and potentialities of that particular place, involving collaboration with various local stakeholders, and combining local sources of funding (e.g., from local business, property or income taxes) with central government or other authority funding streams.
>
> (Martin et al., 2021, p. 89)[2]

Some of the strategies that might be adopted in a place-based approach are:

- Programs to raise the skill levels in left behind places, including the digital skills needed to raise productivity and attract new businesses. The value of education and training is shown by research which found that in the USA the regions with the most lasting negative effects from competition from Chinese imports had lower levels of educational attainment, while those regions with more college graduates were more successful at recovering.
- Programs to raise the productivity and quality of local firms.
- Programs to develop and support local entrepreneurs and the formation of new businesses. This is likely to be more effective than the widespread practice, especially in the USA, of trying to attract new businesses through subsidies and other incentives.
- Building transportation infrastructure to better connect businesses within a region to each other, which will give them more of the benefits of agglomeration, rather than to a major growing city, which may make it easier for the latter to dominate them. Better local public transportation will also make it possible for workers to travel further to find employment.
- Channelling public spending on research and development towards areas with lower productivity. This is the case in Germany, but the UK does the opposite (Stansbury et al., 2023).
- Drawing on people's attachment and sense of belonging to their own place (see Chapter 4) to engage them in local development projects.

Box 11.1 describes an example of a place-based approach.

Box 11.1 Smart Specialisation

An example of a place-based approach is the so-called 'Smart Specialisation' strategy that has been adopted by the European Union and the OECD and implemented in a number of countries. As described by the European Commission, a strategy for smart specialisation is based on five key principles:

- Smart specialisation is a place-based approach, meaning that it builds on the assets and resources available to regions and Member States and on their specific socioeconomic challenges in order to identify unique opportunities for development and growth;
- To have a strategy means to make choices for investment. Member States and regions ought to support only a limited number of well-identified priorities for knowledge-based investments and/or clusters. Specialisation means focusing on competitive strengths and realistic

growth potentials supported by a critical mass of activity and entrepreneurial resources;

- Setting priorities should not be a top-down, picking-the-winner process. It should be an inclusive process of stakeholders' involvement centred on 'entrepreneurial discovery' that is an interactive process in which market forces and the private sector are discovering and producing information about new activities, and the government assesses the outcomes and empowers those actors most capable of realising this potential;
- The strategy should embrace a broad view of innovation, supporting technological as well as practice-based and social innovation. This would allow each region and Member State to shape policy choices according to their unique socioeconomic conditions;
- Finally, a good strategy must include a sound monitoring and evaluation system as well as a revision mechanism for updating the strategic choices. (European Commission, n.d.)

The aim of Smart Specialisation is to build on existing place-based capabilities and extend them through research and innovation. This strategy has been implemented in both urban and rural areas, and targets a wide range of industries. For example, in Andalusia in Spain it is centred on aerospace industries, while in Flanders in Belgium, the focus is on chemistry and chemical industries that can contribute to sustainability. In the rural Gippsland region of the State of Victoria in Australia, on the other hand, the strategy focuses on energy, food and fibre, visitor economy, and health and wellbeing. However, it is unclear whether Smart Specialisation can be effective in all regions and places, and especially in peripheral locations.

To have any chance of success, place-based strategies need to be adequately funded and pursued for a decade or more, as it takes many years to turn around a local or regional economy. Constant changes by governments, whose vision is generally restricted to the next election, must be avoided. To be effective, these strategies also require a decentralisation of power and resources to strengthened local institutions and place-based agencies.

More radical strategies question the meaning of economic growth and the composition of local economies. For example, economists and geographers have questioned the reliance on economic growth, as measured by the value of production, as the indicator for assessing 'development'. Instead, they advocate a greater emphasis on wellbeing and indicators such as health and employment. The focus of many local economic development policies on creating competitive industries that export products and services has also been questioned, for two reasons. One is that this focus is accompanied by a neglect of the 'lower value-added sectors such as social care, hospitality, leisure, public services and retail in which most workers are employed in lagging regions' (MacKinnon et al., 2022, p. 43). The second is that some left behind places may be unable to develop and export high-technology manufactures or advanced business services, a point that was made earlier in relation to Smart Specialisation Strategies.

A related idea is the concept of the 'foundational economy'. This consists of a range of public services, such as health, education, and aged care, and 'activities carried out by private suppliers, such as in personal services, retail, distribution and logistics, construction and maintenance, as well as certain agricultural and food processing activities' (Martin et al., 2022, p. 17). These activities are called 'foundational' because they are the basis of everyday life and need to be located close to the people they serve. They also do not have to compete on export markets, or with suppliers outside their region, and are more stable forms of employment.

How is geographical thinking used in this chapter?

Like the previous chapter, this one has been written to illustrate the application of geographical ways of thinking. As a guide for teachers, the ways in which it does this are summarised in Table 11.4, which lists the core concepts discussed in earlier chapters of the book that are applied in this one.

Table 11.4 How the concepts are applied in the chapter

Core concepts	*Application in the chapter*
Space	A major theme in the chapter is the contrast between national economic policies that are space-blind, and those that recognise spatial variations within the nation.
Place	The chapter discusses regional inequalities, which are inequalities between places, as a geographical perspective on inequality. The national economy is made up of activities in individual places. The political responses of people in left behind places (the geography of discontent) are influenced by conditions in each place. Place-based policies recognise and build on the distinctive characteristics of each place and their different strengths, weaknesses and potentials. They also build on people's attachment to their own place.

(Continued)

Table 11.4 (Continued)

Core concepts	*Application in the chapter*
Interconnection	Interconnection is the underlying concept in theories that expect regional inequalities to decrease through the flow of knowledge from dynamic to lagging places, and of labour in the opposite direction. It is also implicit in explanations of why these theories do not work. Interconnection also underlies the rationale for strategies to improve transportation links between places in lagging regions. The chapter takes a holistic (interconnected) approach to explaining the problems of left behind places.
Environment	Left behind places, because of the industries in which they specialised, have problems of toxic pollution that are a deterrent to new investment.
Human wellbeing	This concept is illustrated in the advocacy for wellbeing as the aim of local economic development, instead of GDP.

Conclusion

Economic geography is generally not a popular part of the school curriculum, except when it is about 'developing countries'. This may be because in the past it was mostly about the location of economic activities, which may have had little interest to students. Yet the economy is a major concern for most people, as measured by public opinion polls, and should have a stronger presence in the school geography curriculum. This chapter describes a different type of economic geography and argues that regional inequality within countries is a significant issue worth studying, and one that might interest students because it is a major current economic, social and political issue. It is also a very geographical topic, and illustrates the contribution of geographers to the understanding of the causes of regional inequalities, and the design of policies to reduce them.

Summary

- Economic inequality is the difference between people in income (their earnings) and wealth (their accumulated assets).
- Income inequality in most high-income countries has increased since around 1980.
- Regional inequality is the difference in people's incomes resulting from where they live, as different places have different industries, occupations, rates of unemployment, and levels of prosperity.

- Regional inequality has increased in many high-income countries.
- Places that have fallen behind the more prosperous regions have been called left behind places.
- There are competing arguments about whether policies should be implemented to improve the economies of these places.
- The political response of left behind places has been called 'a geography of discontent'.
- Additional reasons for reducing inequality are that inequality affects human wellbeing, increases pressures on the environment, reduces national economic growth and has political consequences.
- The causes of places being 'left behind' include technological change, outsourcing of production through international trade, economic policies that favoured cities that were large and dynamic, negative social attitudes and environmental pollution.
- Policies to improve the economies of left behind places and the wellbeing of their populations should be more place-based.

How could you use this chapter in teaching?

Discussion of the causes of regional inequalities, and of policies to address them, is probably best left to the upper secondary years. However, the issue of inequality can be raised in primary school, particularly in relation to inequalities between and within countries, and left behind places can be included in relevant topics anywhere in the secondary years. These are some questions for your students.

- What is economic inequality and how is it measured?
- How does your country compare with other countries in its level of inequality?
- Is your country becoming more, or less, equal?
- What are 'left behind places', and how can they be explained?
- Should governments do more to improve the economies of left behind places and the wellbeing of their populations?
- Why are place-based policies advocated for left behind places?
- Are there any place-based projects in your area, and if there are, investigate how they work?
- Where have the core geographical concepts been used in this topic?
- Why does 'geography matter' in the choice of policies for left behind places?

Notes

1 See also Martin (2008).
2 See also Muro (2023) on US place-based strategies.

Useful reading

General

Economist, Tackling regional inequality in the rich world. https://www.economist.com/films/2023/01/19/tackling-regional-inequality-in-the-rich-world
A short and informative film.

Rodríguez-Pose, A. (2018). The revenge of the places that don't matter (and what to do about it). *Cambridge Journal of Regions, Economy and Society*, *11*, 189–209. https://doi.org/10.1093/cjres/rsx024

On the UK

Martin, R., Gardiner, B., Pike, A., Sunley, P., & Tyler, P. (2021). *Levelling up left behind places: The scale and nature of the economic and policy challenge*. Routledge. https://doi.org/10.4324/9781032244341

McCann, P. (2020). Perceptions of regional inequality and the geography of discontent: Insights from the UK. *Regional Studies*, *54*(2), 256–267. https://doi.org/10.1080/00343404.2019.1619928

On the USA

Hendriksen, C., Muro, M., & Galston, W. A. (2018). *Countering the geography of discontent: Strategies for left-behind places*. Brookings. https://www.brookings.edu/research/countering-the-geography-of-discontent-strategies-for-left-behind-places/

Muro, M. (2023). Biden's big bet on place-based industrial policy. *The Avenue*. https://www.brookings.edu/blog/the-avenue/2023/03/06/bidens-big-bet-on-place-based-industrial-policy/

On Australia

ACTU (Australian Council of Trade Unions). (2018). *Regional inequality in Australia and the future of work*. https://www.actu.org.au/our-work/policies-publications-submissions/2018/regional-inequality-in-australia-and-the-future-of-work

CEDA (Committee for Economic Development of Australia). (2018). *How unequal? Insights on inequality*. CEDA. https://www.ceda.com.au/ResearchAndPolicies/Research/Population/How-unequal-Insights-on-inequality

References

Beatty, C., & Fothergill, S. (2020). Recovery or stagnation? Britain's older industrial towns since the recession. *Regional Studies*, *54*(9), 1238–1249. https://doi.org/10.1080/00343404.2019.1699651

Bez, C., & Virgillito, M. E. (2022). Toxic pollution and labour markets: Uncovering Europe's left-behind places. *LEM Working Paper Series*, No. 2022/19. Scuola Superiore Sant'Anna, Laboratory of Economics and Management (LEM). https://www.lem.sssup.it/WPLem/2022-19.html

Chancel, L., Piketty, T., Saez, E., & Zucman, G. (2022). *World Inequality Report 2022*. World Inequality Lab. https://wir2022.wid.world/

Dabla-Norris, E., Kochhar, K., Ricka, F., Suphaphiphat, N., & Tsounta, E. (2015). Causes and consequences of income inequality: A global perspective. *IMF Staff Discussion Note*, 2015/15/13. https://www.imf.org/en/Publications/Staff-Discussion-Notes/Issues/2016/12/31/Causes-and-Consequences-of-Income-Inequality-A-Global-Perspective-42986

Denti, D., & Faggian, A. (2021). Where do angry birds tweet? Income inequality and online hate in Italy. *Cambridge Journal of Regions, Economy and Society*, *14*, 483–506. https://doi.org/10.1093/cjres/rsab016

Dorling, D. (2015). Income inequality in the UK: Comparisons with five large Western European countries and the USA. *Applied Geography*, *61*, 24–34. https://doi.org/10.1016/j.apgeog.2015.02.004

European Commission. *Smart Specialisation Platform*. https://s3platform.jrc.ec.europa.eu/

Hamann, M., Berry, K., Chaigneau, T., Curry, T., Heilmayr, R., Henriksson, P. J. G., Hentati-Sundberg, J., Jina, A., Lindkvist, E., Lopez-Maldonado, Y., Nieminen, E., Piaggio, M., Qiu, J., Rocha, J. C., Schill, C., Shepon, A., Tilman, A. R., van den Bijgaart, I., & Wu, T. (2018). Inequality and the biosphere. *Annual Review of Environment and Resources*, *43*, 61–83. https://doi.org/10.1146/annurev-environ-102017-025949

Harris, R., & Moffat, J. (2022). The geographical dimension of productivity in Great Britain, 2011–18: The sources of the London productivity advantage. *Regional Studies*, *56*(10), 1713–1728. https://doi.org/10.1080/00343404.2021.2004308

Hill, F. (2021). *There is nothing for you here: Finding opportunity in the twenty-first century*. Mariner Books.

Iammarino, S., Rodríguez-Pose, A, & Storper, M. (2019). Regional inequality in Europe: Evidence, theory and policy implications. *Journal of Economic Geography*, *19*, 273–298. https://doi.org/10.1093/jeg/lby021

Jetten, J., Peters, K., & Casara, B. (2022). Economic inequality and conspiracy theories. *Current Opinion in Psychology*, *47*, 101358. https://doi.org/10.1016/j.copsyc.2022.101358

Los, B., McCann, P., Springford, J., & Thissen, M. (2017). The mismatch between local voting and the local economic consequences of Brexit. *Regional Studies*, *51*(5), 786–799. https://doi.org/10.1080/00343404.2017.1287350

MacKinnon, D., Kempton, L., O'Brien, P., Ormerod, E., Pike, A., & Tomaney, J. (2022). Reframing urban and regional 'development' for 'left behind' places. *Cambridge Journal of Regions, Economy and Society*, *15*, 39–56. https://doi.org/10.1093/cjres/rsab034

Martin, R. (2008). National growth versus spatial equality? A cautionary note on the new 'trade-off' thinking in regional policy discourse. *Regional Science Policy & Practice*, *1*(1), 3–13. https://doi.org/10.1111/j.1757-7802.2008.00003.x

Martin, R., Martinelli, F., & Clifton, J. (2022). Rethinking regional policy in an era of multiple crises. *Cambridge Journal of Regions, Economy and Society*, *15*, 3–21. https://doi.org/10.1093/cjres/rsab037

Martin, R., & Sunley, P. (in press). Capitalism divided? London, financialisation and the UK's spatially unbalanced economy. *Contemporary Social Science* (forthcoming).
McCann, P. (2020). Perceptions of regional inequality and the geography of discontent: Insights from the UK. *Regional Studies*, *54*(2), 256–267. https://doi.org/10.1080/00343404.2019.1619928
Massey, D. (2001). Geography on the agenda 1. *Progress in Human Geography*, *25*(1), 5–17. https://doi.org/10.1191/030913201670520885
Rickard, S. J. (2020). Economic geography, politics and policy. *Annual Review of Political Science*, *23*, 187–202. https://doi.org/10.1146/annurev-polisci-050718-033649
Rodríguez-Pose, A., Storper, M. and Iammarino, S. (2018). *Regional inequality in Europe: Evidence, theory and policy implications.* VOXEU Column. Centre for Economic Policy Research. https://cepr.org/voxeu/columns/regional-inequality-europe-evidence-theory-and-policy-implications
Sayer, A. (2016). *Why we can't afford the rich.* Policy Press.
Soknes, P. E. (2021). *Tomorrow's economy: A guide to creating healthy green growth.* MIT Press.
Stansbury, A., Turner, D., & Balls, E. (2023). *How to tackle the UK's regional economic inequality: Focus on STEM, transport, and innovation.* VOXEU Column, Centre for Economic Policy Research. https://cepr.org/voxeu/columns/how-tackle-uks-regional-economic-inequality-focus-stem-transport-and-innovation
Ulrich-Schad, J. D., & Duncan, C. M. (2018). People and places left behind: Work, culture and politics in the rural United States. *The Journal of Peasant Studies*, *45*(1), 59–79. https://doi.org/10.1080/03066150.2017.1410702
United Nations Department of Economic and Social Affairs. (2020). *World social report 2020: Inequality in a rapidly changing world.* United Nations.
Wilkinson, R., & Pickett, K. (2009). *The spirit level: Why more equal societies almost always do better.* Allen Lane.
Wilkinson, R., & Pickett, K. (2018). *The inner level: How more equal societies reduce stress, restore sanity and improve everyone's well-being.* Allen Lane.

Chapter 12

Conclusion

Geography is a wide-ranging subject, covering a vast variety of topics. This book has argued that what unites it is a group of core concepts. The concepts of place, space, environment and interconnection are ways of thinking about the world that influence the issues geographers choose to study, the questions they ask, the methods of analysis they use and the explanations they investigate. Place, for example, is a way of thinking that can be applied across most, if not all, fields of geography, from fluvial geomorphology to racial segregation. Scale and time influence how geographers investigate, and sustainability and human wellbeing influence the criteria they employ to evaluate what they find.

These functions of the core concepts, and the relationships between them, are illustrated in Figure 12.1. The lines with long dashes represent the influence of the concepts of place, space and environment on the perspectives geographers bring to an investigation, as discussed in the chapters on each concept. They also have an influence on each other, represented by the solid lines between them. Environment, for example, has a major influence on what places are like, while places have an influence on the outcomes of environmental processes because of their unique characteristics. These characteristics are partly influenced by space, because where a place is located influences its environmental and human characteristics. Environments also vary across space, but are interconnected through flows of energy and matter. Interconnection is partly represented by the ellipse around the Topic of study, showing how the influences of the other concepts are integrated in the investigation. It is also represented by the ellipses around place, space and environment, which identify the interrelationships within each of these concepts. The influences of the choice of scale and duration of time on the design of the study, and of the concepts of sustainability and human wellbeing on the evaluation of the results, are represented by the lines with short dashes. This is a highly simplified portrayal of the relationships between the concepts, but the diagram may help students to see an overall pattern into which they can add more detail.

The book has also showed how these concepts support forms of higher order thinking that range from questioning through analysis to generalising, applying and evaluating. But what should students be thinking about? A starting point in looking for an answer might be to find out what young

DOI: 10.4324/9781003376668-13

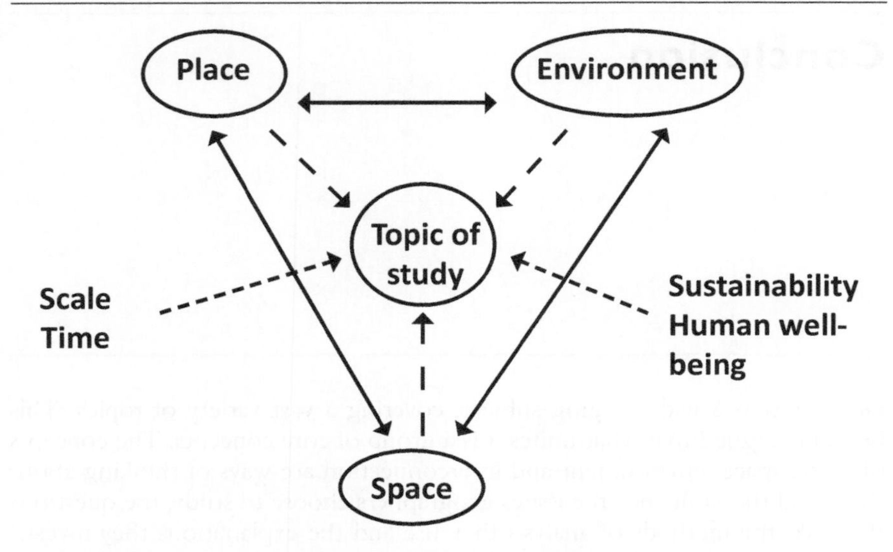

Figure 12.1 The relationships between the core concepts.

people themselves think are the important issues. These are identified in surveys of their opinions conducted in the UK, the USA and Australia in 2022 or 2023.[1] One of these surveys was of youth aged 15–19, and the other two of young people aged from 18 to 24 or 29. In all the surveys, respondents were asked to identify the three most important national issues. The results reveal some differences between countries that might be expected, such as abortion, reproductive health care and gun violence prevention being prominent in the USA and immigration and asylum in the UK. On the other hand, there is considerable agreement across the countries, with the top four most important issues being:

- The economy
- The environment (specified as climate change in the USA)
- Health (specified as health care costs in the USA, and mental health and COVID-19 in Australia)
- Housing and homelessness

Where these issues ranked varied with the characteristics of the respondents. In the USA, for example, 36% of black youth selected racism as one of their three greatest concerns, compared with only 9% of white youth, but black youth were much less likely to be concerned about climate change than white youth. Similarly, in Australia, female respondents were more likely than males to identify the environment, and equity and discrimination, as top concerns. Findings such as these show that it is important to identify the concerns and interests of different sections of the population, including minority and marginalised groups.

Overall, however, the economy and the environment were the equal top most important issues across the three countries, followed by health and housing. The environment is well represented in school geography, although this may not always include the areas that most concern young people. Climate change is an essential topic for study but needs to be examined geographically, and Chapter 10 described one way to do this. Chapter 6 discussed other areas of physical geography that are vital to an understanding of the ways that humans depend on the environment that tend to be neglected in school geography, and are needed for students to understand many of the changes that are part of the Anthropocene. The other three concepts are not as well represented in school geography as the environment but are topics that perhaps should be promoted. Chapter 11 showed that regional economic inequality is a major problem within nations and not just between them, and that geographical thinking has much to contribute to understanding its causes and consequences, and how regional inequalities might be reduced. Health has been discussed in several chapters, and is a growing area of geographical research that students find interesting. Housing is the least likely of the four issues to be included in school geography, except perhaps in relation to urbanisation in lower-income countries, yet some of the leading housing researchers have come from geography, and it is an issue of growing concern to young people.

Whatever topics and issues teachers choose for their students, I hope that they will use them to teach young people to think geographically, deeply and ethically, because showing ways to do this, and to make geography a challenging subject for young thinkers, has been the purpose of the book.

Note

1 The surveys were:
   (a) UK: YouGov survey of people aged 18–24, on 10.04.2023. The question asked was: Which of the following do you think are the most important issues facing the country at this time? Please tick up to three. https://yougov.co.uk/topics/education/trackers/the-most-important-issues-facing-the-country.
   (b) US: Youth survey by Center for Information & Research on Civic Learning and Engagement of people aged 18–29, showing the percentage who named each issue as one of their top 3 concerns. https://circle.tufts.edu/latest-research/youth-2022-concerned-about-issues-neglected-campaigns.
   (c) Australia: Mission Australia, Youth Survey 2022. The question asked people aged 15–19 what are the three most important issues in Australia. https://www.missionaustralia.com.au/publications/youth-survey/2618-youth-survey-2022-report/file.

Index

Note: Page numbers followed by "n" refer to end notes.